Lecture Notes in Computer Science 14891

Founding Editors

Gerhard Goos
Juris Hartmanis

Editorial Board Members

Elisa Bertino, *Purdue University, West Lafayette, IN, USA*
Wen Gao, *Peking University, Beijing, China*
Bernhard Steffen, *TU Dortmund University, Dortmund, Germany*
Moti Yung, *Columbia University, New York, NY, USA*

The series Lecture Notes in Computer Science (LNCS), including its subseries Lecture Notes in Artificial Intelligence (LNAI) and Lecture Notes in Bioinformatics (LNBI), has established itself as a medium for the publication of new developments in computer science and information technology research, teaching, and education.

LNCS enjoys close cooperation with the computer science R & D community, the series counts many renowned academics among its volume editors and paper authors, and collaborates with prestigious societies. Its mission is to serve this international community by providing an invaluable service, mainly focused on the publication of conference and workshop proceedings and postproceedings. LNCS commenced publication in 1973.

Marius Rohde Johannessen · Csaba Csáki ·
Lieselot Danneels · Sara Hofmann ·
Thomas Lampoltshammer · Peter Parycek ·
Gerhard Schwabe · Efthimios Tambouris ·
Jolien Ubacht
Editors

Electronic Participation

16th IFIP WG 8.5 International Conference, ePart 2024
Ghent, Belgium, September 3–5, 2024
Proceedings

Editors
Marius Rohde Johannessen
University of South-Eastern Norway
Borre, Norway

Lieselot Danneels
Ghent University
Ghent, Belgium

Thomas Lampoltshammer
University for Continuing Education Krems
Krems, Austria

Gerhard Schwabe
Zürich University
Zurich, Switzerland

Jolien Ubacht
Technical University Delft
Delft, The Netherlands

Csaba Csáki
Corvinus University of Budapest
Budapest, Hungary

Sara Hofmann
University of Agder
Kristiansand, Norway

Peter Parycek
University for Continuing Education Krems
Krems, Austria

Efthimios Tambouris
University of Macedonia
Thessaloniki, Greece

ISSN 0302-9743 ISSN 1611-3349 (electronic)
Lecture Notes in Computer Science
ISBN 978-3-031-70803-9 ISBN 978-3-031-70804-6 (eBook)
https://doi.org/10.1007/978-3-031-70804-6

© IFIP International Federation for Information Processing 2024

This work is subject to copyright. All rights are solely and exclusively licensed by the Publisher, whether the whole or part of the material is concerned, specifically the rights of translation, reprinting, reuse of illustrations, recitation, broadcasting, reproduction on microfilms or in any other physical way, and transmission or information storage and retrieval, electronic adaptation, computer software, or by similar or dissimilar methodology now known or hereafter developed.
The use of general descriptive names, registered names, trademarks, service marks, etc. in this publication does not imply, even in the absence of a specific statement, that such names are exempt from the relevant protective laws and regulations and therefore free for general use.
The publisher, the authors and the editors are safe to assume that the advice and information in this book are believed to be true and accurate at the date of publication. Neither the publisher nor the authors or the editors give a warranty, expressed or implied, with respect to the material contained herein or for any errors or omissions that may have been made. The publisher remains neutral with regard to jurisdictional claims in published maps and institutional affiliations.

This Springer imprint is published by the registered company Springer Nature Switzerland AG
The registered company address is: Gewerbestrasse 11, 6330 Cham, Switzerland

If disposing of this product, please recycle the paper.

Preface

The EGOV-CeDEM-ePart 2024 conference, for short EGOV 2024, was organized by the IFIP WG 8.5 on ICT and public administration. The conference is dedicated to digital or electronic government, open government, local government (smart cities), smart governance, artificial intelligence (AI), e-democracy, policy informatics, and electronic participation. With a long tradition along its various branches, the EGOV-CeDEM-ePart conference has gained its reputation of being the key and leading conference worldwide in the research domains of digital/electronic, open, and smart government, as well as electronic participation.

The call for papers attracted completed research papers, work-in-progress papers on ongoing research (including doctoral papers), project and case descriptions, as well as workshops and panel proposals. This volume contains only full papers. All submissions were assessed through a double-blind peer-review process, with at least three reviewers per submission, and the acceptance rate was 42%. The submitted papers were of high quality, and many reviewers indicated that the topic of the papers they reviewed were novel.

The review process was based on a double-blind reviewing process and conflicts of interest were actively avoided. Authors of papers submitted their papers to a track. The track chairs only had access to the papers within their own track and could assign reviewers and propose acceptance decisions. The chairs of the tracks were editors of the proceedings, in addition to the general chairs. Track chairs were not allowed to submit to their own track, nor were their close collaborators, to avoid any conflict of interest. Track chairs could either submit to another track or to the 'track chairs' track. The latter was handled by one of the general chairs, and these submissions were reviewed in the same way as all other submissions. The general chairs checked after the submission deadline whether there was any conflict of interest among the papers submitted to tracks, and if so, the respective paper was moved to another track. The track chairs checked that all papers were submitted anonymously. If not, the authors were asked to resubmit within a couple of days. Track chairs assigned the reviewers and selected the programme committee members in such a way that experts in this area reviewed the papers and there were no conflicts of interest. After at least three reviews were received, the track chairs made a proposal for a decision per paper. The decisions were discussed in a meeting with the general and track chairs to ensure that the decisions were made in a consistent manner per track.

The conference tracks of the 2024 edition present the evolution of the topics and the progress in this field. The papers in these proceedings were distributed over the following tracks:

- General E-Government and E-Governance
- Emerging Issues and Innovations
- Smart Cities (Government, Districts, Communities, and Regions)
- AI, Data Analytics, and Automated Decision Making

– Open Data: Social and Technical Aspects

Among the full research paper submissions, 29 papers (empirical and conceptual) were accepted for this year's Springer LNCS EGOV proceedings (vol. 14841). The LNCS ePart proceedings (LNCS vol. 14891) contain the 15 completed research papers from the following tracks:

– General E-Democracy and E-participation
– Digital Technologies and Sustainable Development
– Governance, Digital Legislation, and Policy
– Digital and Social Media
– Digital Society; Artificial Intelligence Governance and its Societal Challenges

The accepted ongoing work, posters, and practitioners' papers are published in the Joint Proceedings of Ongoing Research, Practitioners, Posters, Workshops, and Projects at the EGOV-CeDEM-ePart 2024 conference. The CEUR Workshop Proceedings (CEUR-WS.org) are free, open-access proceedings for Computer Science Workshops.

As in the previous years and per the recommendation of the Paper Awards Committee under the leadership of Manuel Pedro Rodríguez Bolívar of the University of Granada, Spain, the IFIP EGOV-CeDEM-ePart 2024 conference granted outstanding paper awards in three distinct categories:

– The most interdisciplinary and innovative research contribution
– The most compelling critical research reflection
– The most promising practical concept

The winners in each category were announced during the awards ceremony at the conference.

For the first time, a Junior Faculty School was organized to connect PhD students, postdoctoral students, and junior faculty to the e-government research community to provide advice on how to plan and manage the tenure process and their future career, and to offer a platform for discussions about theory and methods, as well as networking. The Junior Faculty School consisted of keynotes, panels, discussions, and interactive sessions designed for junior faculty to be oriented and prepared for the opportunities and challenges ahead.

The IFIP EGOV 2024 conference was hosted in Belgium, in the heart of Europe, by Ghent University and KU Leuven. They are both top 100 universities and two of the major universities in Belgium with over 50,000 students and 15,000 employees. Ghent University was founded in 1817 and KU Leuven in 1425. In addition to all that history, the universities also bring a lot of innovation to their vibrant towns. As part of the conference, the University of Ghent hosted the PhD colloquium and Junior Faculty School in Ghent on 1–2 September. The PhD Colloquium was a full-day event on 1 September, and offered a limited number of PhD bursaries. The Junior Faculty School was a full-day event on 2 September, for sharing experiences to advance the career of young researchers interested in digitalization and government. The conference sessions were hosted from 3–5 September in Leuven at KU Leuven.

Many people behind the scenes make large events like this conference happen. We would like to thank the members of the Program Committee, the reviewers, and the track chairs for their great efforts in reviewing the submitted papers. We would also like to express our deep gratitude to Joep Crompvoets and his local team at KU Leuven, and Lieselot Danneels and her local team at Ghent University, for hosting the conference.

The papers in these proceedings range from societal to scientific, and from rigorous to innovative. We hope you enjoy reading them!

September 2024

Marius Rohde Johannessen
Csaba Csáki
Lieselot Danneels
Sara Hofmann
Thomas Lampoltshammer
Peter Parycek
Gerhard Schwabe
Efthimios Tambouris
Jolien Ubacht

Organization

Program Committee

Wirawan Agahari	TU Delft, Netherlands
Valerie Albrecht	Danube University Krems, Austria
Laura Alcaide-Muñoz	University of Granada, Spain
Dwayne Ansah	Utrecht University, Copernicus Institute of Sustainable Development, Netherlands
Leonidas Anthopoulos	University of Thessaly, Greece
Frank Bannister	Trinity College Dublin, Ireland
Ana Alice Baptista	University of Minho, Portugal
Peter Bellström	Karlstad University, Sweden
Flavia Bernardini	Universidade Federal Fluminense, Brazil
Lasse Berntzen	University of South-Eastern Norway, Norway
Nitesh Bharosa	TU Delft, Netherlands
Christina Bidmon	Utrecht University, Netherlands
Radomir Bolgov	Saint Petersburg State University, Russia
Nicolas Bono Rossello	Namur Digital Institute, University of Namur, Belgium
Alessio Bramini	University of Tuscia, Italy
Karin Brodén Ahlin	Karlstad University, Sweden
Iván Cantador	Universidad Autónoma de Madrid, Spain
Luiz Paulo Carvalho	Federal University of Rio de Janeiro, Brazil
Cesar Casiano Flores	University of Twente, Netherlands
Seong Wook Chae	Hoseo University, South Korea
Youngseok Choi	Kingston University London, UK
Vincenzo Ciancia	Istituto di Scienza e Tecnologie dell'Informazione, Consiglio Nazionale delle Ricerche, Pisa, Italy
Antoine Clarinval	Université de Namur, Belgium
Rodrigo Conde	Austrian Institute of Technology, Austria
Andreiwid Sheffer Corrêa	Federal Institute of São Paulo, Brazil
MaríaElicia Cortés-Cediel	Universidad Complutense de Madrid, Spain
J. Ignacio Criado	Universidad Autónoma de Madrid, Spain
Joep Crompvoets	KU Leuven-Public Governance Institute, Belgium
Jonathan Crusoe	Gothenburg University and University of Borås, Sweden
Csaba Csaki	Corvinus University - Corvinus Business School, Hungary

Frank Danielsen	University of Agder, Norway
Lieselot Danneels	Ghent University/Vlerick Business School, Belgium
Simon Dechamps	Namur Digital Institute, UNamur, Belgium
Devin Diran	TNO Vector, Netherlands
Bettina Distel	Universität Münster, Germany
Roel Dobbe	Delft University of Technology, Netherlands
David Duenas-Cid	Kozminski University, Poland
Matthias Döring	University of Southern Denmark, Denmark
Noella Edelmann	Danube University Krems, Austria
Claudia Egher	Utrecht University, Netherlands
Gregor Eibl	Danube University Krems, Austria
Ali El Samra	Brunel University London, UK
Eiri Elvestad	University of South-Eastern Norway, Norway
Loukis Euripides	University of the Aegean, Greece
Kalampokis Evangelos	University of Macedonia, Greece
Claucia Faganello	PUCRS, Brazil
Joan-Francesc Fondevila-Gascón	CECABLE, UdG (Escola Universitària Mediterrani) and UAB, Spain
Sabrina Franceschini	Regione Emilia-Romagna, Italy
Alizée Francey	University of Lausanne, Switzerland
Mary Francoli	Carleton University, Canada
Asbjørn Følstad	SINTEF, Norway
Larissa Galdino de	Magalhães Santos United Nations University (UNU-EGOV), Portugal
Francisco García Morán	European Commission, Luxembourg
Mila Gasco-Hernandez	University at Albany, SUNY, USA
Alexandros Gerontas	University of Macedonia, Greece
Katarina Gidlund	Mid Sweden University, Sweden
Ramon Gil-Garcia	University at Albany, State University of New York, USA
Dimitris Gouscos	University of Athens, Greece
Divya Kirti Gupta	Indus Business Academy, India
Yash Gupta	University of Southern California, USA
Mariana Gustafsson	Linköping University, Sweden
Tessa Haesevoets	Ghent University, Belgium
Yang Hai	Brunel University London, UK
Annika Hasselblad	Mid Sweden University, Sweden
Marcus Heidlund	Mid Sweden University, Sweden
Natali Helberger	University of Amsterdam, Netherlands
Sara Hofmann	University of Agder, Norway
Roumiana Ilieva	Technical University of Sofia, Bulgaria

Tomasz Janowski	Gdańsk University of Technology, Poland
Marijn Janssen	Delft University of Technology, Netherlands
Soohyun Jeon	Brunel University London, UK
Carlos E. Jimenez-Gomez	National Center for State Courts, USA
Marius Rohde Johannessen	University of South-Eastern Norway, Norway
Gustaf Juell-Skielse	University of Borås, Sweden
Muneo Kaigo	University of Tsukuba, Japan
Evangelos Kalampokis	University of Macedonia, Greece
Nikos Karacapilidis	University of Patras, Greece
Areti Karamanou	University of Macedonia, Greece
Naci Karkin	Pamukkale University, Turkey
Singara Rao Karna	Centre for Electronic Government and Open Democracy, Sydney-Melbourne, Australia
Junchul Kim	Brunel University London, UK
Fabian Kirstein	Fraunhofer FOKUS, Germany
Bram Klievink	Leiden University, Netherlands
Agnieszka Kochanowicz	University for Continuing Education Krems, Austria
Zsófia Kräussl	Bayes Business School, City University of London, UK
Thomas Lampoltshammer	University for Continuing Education Krems, Austria
Beatriz Barreto Brasileiro Lanza	CTG SUNY & IDB, USA
Habin Lee	Brunel University London, UK
Hongjoo Lee	Catholic University of Korea, South Korea
Ida Lindgren	Linköping University, Sweden
Johan Linåker	RISE Research Institutes of Sweden, Sweden
Euripides Loukis	University of the Aegean, Greece
Luis F. Luna-Reyes	University at Albany, SUNY, USA
Marie Anne Macadar	Federal University of Rio de Janeiro, Brazil
John McNutt	University of Delaware, USA
Ulf Melin	Linköping University, Sweden
Sehl Mellouli	Université Laval, Canada
Mara Mendes	University of Münster, Germany
Tobias Mettler	University of Lausanne, Switzerland
Edimara Mezzomo Luciano	PUCRS, Brazil
Harekrishna Misra	Institute of Rural Management Anand, India
Gianluca Misuraca	gfa-group, Germany
José María Moreno-Jiménez	Universidad de Zaragoza, Spain
Francesca Morselli	Delft University of Technology, Netherlands
Francesco Mureddu	Lisbon Council, Belgium
Anastasija Nikiforova	University of Tartu, Estonia

Sem Nouws	Delft University of Technology, Netherlands
Anna-Sophie Novak	Danube University Krems, Austria
Niclas Olofsson	Mid Sweden University, Sweden
Monica Palmirani	CIRSFID, ALMA-AI, Italy
Panos Panagiotopoulos	Queen Mary University of London, UK
Laura Piscicelli	Utrecht University, Netherlands
Blaž Podgorelec	Graz University of Technology, Austria
Anuj Puri	Tilburg University, Netherlands
András Pünkösty	Ludovika University of Public Service, Hungary
Aya Rizk	Linköping University, Sweden
Nina Rizun	Gdańsk University of Technology, Poland
Manuel Pedro Rodríguez Bolívar	University of Granada, Spain
Lucas Roldan	PUCRS, Brazil
Alexander Ronzhyn	University of Koblenz-Landau, Germany
Boriana Rukanova	Delft University of Technology, Netherlands
Michael Räckers	European Research Center for Information Systems, Germany
Rodrigo Sandoval-Almazan	Universidad Autónoma del Estado de Mexico, México
Diogo Sasdelli	Donau-Universität Krems, Austria
Antonia Sattlegger	Delft University of Technology, Netherlands
Carsten Schmidt	University of Tartu, Estonia
Hendrik Scholta	University of Münster/ERCIS, Germany
Johannes Scholz	Graz University of Technology, Austria
Uwe Serdült	Ritsumeikan University, Japan
Masoud Shahmanzari	Brunel University London, UK
Tobias Siebenlist	Rhine-Waal University of Applied Sciences, Germany
Anthony Simonofski	University of Namur, Belgium
Koen Smit	University of Applied Sciences Utrecht, Netherlands
Carl-Johan Sommar	Linköping University, Sweden
Reni Sulastri	Delft University of Technology, Netherlands
Leif Sundberg	Mid Sweden University, Sweden
Iryna Susha	University of Utrecht, Netherlands
Efthimios Tambouris	University of Macedonia, Greece
Evrim Tan	KU Leuven Public Governance Institute, Belgium
Luca Tangi	Joint Research Centre-European Commission, Italy
Carolina Tavares Lopes	Pontifical Catholic University of Rio Grande do Sul (PUCRS), Brazil
Lörinc Thurnay	University for Continuning Education Krems, Austria

Jolien Ubacht	Delft University of Technology, Netherlands
Floris van Krimpen	Delft University of Technology, Netherlands
Michael Veale	University College London, UK
Marco Velicogna	IRSIG-CNR, Italy
Gabriela Viale Pereira	Danube University Krems, Austria
Shefali Virkar	WU Vienna University of Economics and Business, Austria
Gianluigi Viscusi	Linköping University, Sweden
Jörn von Lucke	Zeppelin Universität Friedrichshafen, Germany
Bianca Wentzel	Fraunhofer FOKUS, Germany
Elin Wihlborg	Linköping University, Sweden
Mete Yildiz	Hacettepe Üniversitesi İİBF, Turkey
Maija Ylinen	Tampere University, Finland
Thomas Zefferer	A-SIT Plus GmbH, Austria
Dimitris Zeginis	Centre for Research and Technology Hellas, Greece
Sheila Zimic	Association of Local Authorities in Västernorrland County, Sweden
Anneke Zuiderwijk	Delft University of Technology, Netherlands

Contents

Are Children Ready for the Metaverse? The Minefield of Virtual
Participation in Digital Social Spaces with Harmful Content and Behavior 1
   Robin Effing and Michael Hinz

Navigating Privacy Regulations: Administrative Burden of Digital
Self-Services for Vulnerable Citizens 16
   Ida Heggertveit

Mitigating Administrative Burdens: Understanding the Role
of Intermediaries in Co-producing Digital Self-services 31
   *Hanne Höglund Rydén, Sara Hofmann,
Luiz Henrique Alonso de Andrade, and Ida Heggertveit*

Getting Rid of It: An Unlearning Perspective on Digital Government
Competences 47
   *Michael Koddebusch, Bettina Distel, Marco Di Maria, Paul Brützke,
and Jörg Becker*

Public Management Competencies in a Digital World: Lessons
from a Global Frontrunner 64
   Ulrik B. U. Roehl and Joep Crompvoets

For the Fun of IT– In Search for Sensemaking of Digitalization Training
Program for Leaders in Schools and Pre-schools in Sweden 83
   Elin Wihlborg and Maria Spante

E-Participation Without Democracy: Understanding Variation in Digital
Engagement in Non-democracies 99
   Thomas Hayes and Martin Karlsson

Sustainable eParticipation Through Lightweight Democracy? 116
   Marius Rohde Johannessen, Noella Edelmann, and Lasse Berntzen

Ethical Governance of Emerging Digital Technologies in the Public
Sector: Insights from Dutch Digital Ethics Commissions 131
   Antonia Sattlegger

Data Pollution: Definition and Policy Responses 147
   Leonardo Mori, Alizée Francey, and Tobias Mettler

Understanding the Problem Space for Effective Use of a Circular Economy
Monitor in Policy Making 163
   Michiel Pauwels, René Reich, An Vercalsteren, Maarten Christis,
   Luc Alaerts, and Karel van Acker

Exploring Data Altruism as Data Donation: A Review of Concepts, Actors
and Objectives 179
   Dwayne Ansah and Iryna Susha

A Method for the Collaborative and Semi-automated Generation
of Conceptual Models from Legal Regulations in Public Organizations 194
   Binh An Patrick Nguyen and Hendrik Scholta

Transparency in Open Government Data Portals: An Assessment of Web
Tracking Practices Across Europe 209
   Stefan Stepanovic, Leonardo Mori, Alizée Francey, and Tobias Mettler

Understanding Trust Frameworks: Goals and Components Identified
Through a Case Study 223
   Louise van der Peet, Nitesh Bharosa, Sander Dijkhuis,
   and Marijn Janssen

Author Index 239

Are Children Ready for the Metaverse? The Minefield of Virtual Participation in Digital Social Spaces with Harmful Content and Behavior

Robin Effing[✉] and Michael Hinz

University of Twente, P.O. Box 217, Enschede 7500 AE, The Netherlands
r.effing@utwente.nl

Abstract. The technological concept of the metaverse has the potential to change our society to an extent that can be compared with past inventions such as the internet. The risks and challenges it might bring for its users when participating are often neglected. Past experiences regarding social media or online gaming showed how such technologies can have negative effects on the psychological and physiological well-being of their users and that malicious individuals might abuse them. A user segment that is especially vulnerable in this context is the group of children and adolescents-youth. This study contains a detailed content analysis of two different metaverse applications regarding what types of child-inappropriate content or behavior exist there and reveals a number of serious concerns when children enter such digital virtual spaces. This paper addresses the types risks the use of the metaverse can pose for children and adolescents by creating a knowledge base derived from the current state of literature and interviews with experts. Additionally, it contributes to research by developing a classification scheme that helps to identify and categorize child-inappropriate content or behavior that occurs in metaverse spaces.

Keywords: Metaverse · digital ethics · virtual reality · augmented reality · social media · VRChat · Recroom · electronic participation Track: Digital Society

1 Introduction

The emergence of the concept metaverse, a virtual world existing as a digital twin in the digital space, will create unprecedented opportunities as well as challenges for society. This development is greatly facilitated by revolutionary technologies such as "virtual reality (VR) which has been aided by the convergence of digital advancements in AI, Internet of Things, Cloud computing, Big Data, and other technologies" [1]. These virtual spaces can provide a multitude of functions ranging from simple aspects such as gaming or communication up to the creation of effective workspaces or digital-twin models of real-life scenarios (e.g. city planning or natural catastrophe prevention) [1]. This range alone shows how diverse and universally applicable the concept of the metaverse will become in the upcoming years, which indicates the importance for not only special interest groups but also for all members of society. The general concept of the metaverse was introduced by Neal Stephenson in his novel "Snow Crash". In this book he depicted a dystopian reality of our world in which people entered a virtual space via a head-mounted display (HMD) where they were able to construct their own personalized world according to their preferences [2, 3]. Inspired by Neal Stephenson's visionary concept, the metaverse is gradually taking shape with technological advancements like "Second Life" or Meta's recent venture "Horizon Worlds" [4]. These immersive spaces allow users to enter digital environments in which they have the possibility to interact via customizable digital representations of themselves (Avatars), bringing us closer to Stephenson's vision [5]. As depicted in Damar [6] a possible and inclusive definition of the metaverse could be a "shared 3D virtual world in which all activities can take place using augmented and virtual reality equipment". A universal metaverse will provide future society with tremendous opportunities for taking part in a digital society based on 3d interaction. This metaverse will reshape digital interactions between people and the way they play, learn, work and exercise. On top of that, new unprecedented ways for entertainment, adventure and fantasy are presented in an advanced 3d interface.

With the metaverse's growing popularity ethical concerns naturally were raised simultaneously [5]. Currently, the development and regulations of certain metaverse spaces are predominantly in the hands of big tech companies. These exist largely outside of the control of respective government agencies. These big tech companies are not fully transparent about their interests and could follow underlying motives which are to a large extent commercial and sometimes even political. Given these hidden interests, the ethical accountability and transparency regarding the regulation of the metaverse are highly questionable [7]. Especially children and young adolescents are particularly endangered by such influences. The mental health of children could be at stake since the metaverse will employ similar content-filtering algorithms as being used within social media and its addictive life logging behavior. Furthermore, the metaverse build upon the gamification principles being applied from the experience of 3d gaming. The dangers created by increased utilization of the metaverse and related technologies could include for example the publication of private data, addiction, increased violent behavior, negative impacts on psychological well-being, cyberbullying, or exposure to unsafe content or behavior that is not appropriate for children and adolescents [8–12]. Due to the unknown impact, the metaverse will have on our society, and the lack of regulations currently defining those virtual spaces, the subject of child and adolescent protection will represent an essential future research domain. Despite the ethical importance of user protection and the prevention of aspects such as harassment or the distribution of child-inappropriate content or behavior, the current state of research regarding the metaverse is predominantly focused on technical issues and the potential positive impacts this new technology might have on our society. Aspects like the underlying technological processes, future business and monetarization possibilities, improvements regarding city planning and urbanization, or advantages for health care and education currently represent the core of literature [1, 2, 6, 7, 13, 14]. Only a fraction of research examines what negative effects a further implementation of the metaverse in our daily life might encompass on young people's health and wellbeing. Considering the novel nature of this metaverse-like technology, it is obvious that long-term studies regarding impacts on our well-being still do not yet exist. Due to this clear gap in research, the aim of this paper is the provision of an initial knowledge base regarding what psychological and physiological risks exist regarding entering these virtual worlds. Furthermore, this guides the development of a conceptual framework used in answering this paper's research question: "What types of child-inappropriate content or behavior can be identified in social or gaming spaces of the metaverse that children and adolescents might encounter during their play?". The remainder of this paper is structured as follows. First, we will show a contextual background based on both literature review and preliminary interview findings resulting in a theoretical framework. In the method chapter we elaborate upon the design of the study. Consequently, we will show the results and finalize the paper with the discussion and conclusion.

2 Theoretical Background

The following section will provide an overview of how the current state of research examined predominant risks the usage of the metaverse could pose for children and adolescents. The literature review took place on the interdisciplinary research databases "Scopus" and "Web of Science" via an initial keyword search of relevant concepts based on a Boolean search strategy. This was then combined with a snowballing technique in which citation tracking was utilized to further discover relevant papers. The concept of the metaverse can be seen as the next developmental step in which technologies such as gaming and social media will take place [10–12, 15]. This paper will follow this approach and takes research regarding relevant psychological and physiological risks of social media and gaming into account that are comparable or potentially generalizable for aspects of the metaverse.

Addiction Risks

There are certain concepts that the metaverse and the closely related VR technology incorporate to a greater extent than traditional comparable technologies such as desktop computing or the use of smartphones. Metaverse and VR are capable of increasingly fostering the occurrence of addictive behavior or other health risks that will be mentioned in the following sections. These are namely the factors of immersion and its closely connected concepts of presence, absorption, and embodiment [9, 16–18]. Immersion defines the extent to which users of a certain technology feel present in a virtual environment which subsequently increases the chance of blurring the borders between virtual reality and real life [5, 17, 19]. Why this factor can have a higher impact on children and adolescents playing violent video games becomes evident when taking the super-realistic properties into account that the metaverse might bring [14]. Lee et al. [10] argue that this might lead to users attempting something they experienced during their play in real life, possibly even immoral behavior. Given the evidence that younger people prefer violent aspects of games more than adults do and that additionally, studies found significant relationships between video game violence and real-life aggression one can argue that exposure of minors to virtual reality violence (amplified by a high level of immersion) creates a concerning risk for society [20, 21]. This compelling and realistic sensory experience for users can be created through technological and design elements that can lead to immersive visual, auditory, and haptic stimuli as well as realistic interaction mechanisms [10]. Furthermore, Cairns et al. [16] mention that a person who is engaged in an immersive state might try to remain in this world, which originates from the self-sustaining characteristic of immersion and makes them hesitant to leave [1, 5, 8]. In case a sufficient degree of immersion is reached, the effect can occur that users feel present in a virtual environment even though they are not physically there [17, 22]. This factor is essential for software and game developers to take into account since a great level of immersion and presence can lead to a stronger engagement and emotional connectedness to the virtual world [9]. Furthermore, Turel et al. [23] provided a study that linked an addiction of children to information systems to the potential development of cardio-metabolic disturbances. As mentioned, bizarre sleeping patterns and insomnia are common themes related to the higher use of digital media due to their emission of short wavelength light which subsequently has an altering impact on the human's production of the sleep hormone melatonin, an effect that is even increased for children [24, 25]. Since most of the studies regarding pre-conditions or effects of addiction are cross-sectional, reverse causation of the mentioned effects is possible.

Risks Through Embodying Avatars
An avatar can be defined as the digital representation of an individual in the metaverse [10]. As the user creates this avatar, possibilities rise to freely customize certain attributes such as the virtual appearance. This ability to customize one's physical attributes and to embody a character that potentially mirrors one's profound wishes for one's appearance in real life creates a highly addictive potential while fostering the blurring of borders between virtual reality and real life [19]: "As the avatar grows, the user's intimacy increases, becoming more immersed in the metaverse. However, addiction and excessive immersion result in confusion and lack of interest in the incongruity with the real world" [5]. If this addiction to the avatar becomes stronger, and the user might start preferring this representation over being themselves, negligence of one's own body and the real world might occur [26]. Young people might be increasingly at risk here due to their tendency to favor products that fulfill their role projections and fantasy [27]. Another aspect is based on the individual's dynamics of behavior [10]. Multiple research papers mentioned the Proteus effect in connection to how the embodiment of virtual avatars can influence certain characteristics or behavioral traits in real life [10, 12, 28–30]. Fox et al. [28] define the Proteus effect as a state of "when a user's self-representation is modified in a meaningful way that is often dissimilar to the physical self." The results of that study were an increase in body-related thoughts, rape-myth acceptance, and body dysmorphia among young women that embodied sexualized avatars. This aligns with Usmani et al. [12], who mention a potential connection between the Proteus effect and the development of body dysmorphia or other related mental illnesses such as anorexia. Absorption can be seen as closely related to immersion as both are defined by an increased mental focus and emotional investment [9, 18, 31]. For the aspect of absorption, a study was conducted by Lavoie et al. [9] that indicated that the degree of absorption has a mediating effect on the relationship between VR experiences and negative emotions. A high degree of

embodiment has a positive effect on the overall sense of immersion and presence the user feels [18].

Risks Delivered by Harmful Intentions
While different types of online harassment and their effects are already identified in research, experts indicated that these occurrences simultaneously take place in online virtual reality as well [32–35]. Blackwell et al. [32] conducted a study that shows that harassment, coupled with the immersive VR technology associated with the metaverse, can increase the psychological impact on the user's mental health compared to traditional online spaces. Firstly, there is verbal harassment and includes aspects such as personal insults, hate speech, or sexualized language. Secondly, physical harassment could take place referring to unwanted touching, standing too close to the user, obstructing the user's movement, or performing visible sexual gestures. Lastly, environmental harassment consists of displaying sexual/violent content, drawing sexual images, or throwing objects at other users. Participants of the by Blackwell et al. [32] conducted interviews indicated that the psychological impact they felt after the harassment took place in virtual reality was greater compared to other online mediums, such as social media. Since VR creates the feeling of being physically present in this space, physical altercations or harassment can induce a similar level of realness as in real life [17]. Jonsson et al. [36] identified poor self-esteem and a poor relationship with parents as risk factors that lead adolescents to search for validation in online spaces and potentially become victims of sexual predators. The risk of grooming children and adolescents in the metaverse by sexual predators describes only partially what kind of threats other users could impose on them. Research conducted by the CCDH provides evidence regarding numerous child inappropriate content or behavior in one of the most popular metaverse social spaces, namely "VRChat" [37]. Besides the already indicated risks of harassment and exposure to explicit content, the study mentioned grooming of children to repeat racist slurs and extremist talking points. Terrorist groups or other extremist parties could use the metaverse to groom children for their cause and provide training opportunities [38, 39].

3 Method

The method of this qualitative study is twofold. First, we conducted a series of exploratory interviews with field experts. Since no long-term studies regarding risks of the anticipated metaverse existed yet, interviewees can only speak from their experience and expertise from related study fields. Given the novel nature of the field, the interviews were quite open and the analysis had the primary goal to develop a preliminary classification framework for the systematic observation that followed. The interview results were transcribed, analyzed and clustered aiming at extending our view. Second, there was a systematic observation of risks in selected applications that resemble the metaverse. This study contains a detailed content analysis of two different metaverse applications regarding the existence of child-inappropriate content and behavior. We followed the rules from Popper [45] to ensure us to provide inquiring evidence that follow the rules of logic. For example, only observing a limited number of cases of harm ensures the existence of such harm. However, we make no claims whatsoever regarding the extent to which these findings are generalizable. Due to the lack of research in this field an exploratory study needed to be conducted to create a knowledge basis and recommendations for future research. Even though the universal metaverse did not exist yet at the time of writing, there are currently applications available that can be seen as social virtual spaces that have already many of the characteristics of the anticipated metaverse. We selected two applications for our systematic observation of potential risks of children in the future metaverse: VRChat and RecRoom. Both applications were chosen under the criteria that they are free-to-play multiplayer spaces, factors that lead to higher user numbers, and subsequently to an increased amount of potentially inappropriate users and occurrences that may cause harm for minors. These spaces can be seen as predecessors to what the metaverse is aiming to be since most of its predicted and aspired functions are still in their infancy so it would be an exaggeration to present them as a viable mirror world to our real life [1]. However, future metaverse worlds could include augmented reality functions and a more advanced digital twin reality which are currently not part of the design of these selected applications.

The data collection method was carried out as follows. We performed a systematic observation inside the selected metaverse applications for a duration of three hours each. We are following a similar approach as the exploratory study conducted by the CCDH [37]. To ensure reliability, the three hours were distributed over multiple times in a day (morning, noon, and night). The maximum duration for which we entered the metaverse was limited to one hour per day. We did not communicate with other users in any way in order to prevent any diversion caused by e.g., gender or age. Combined, this approach averts the occurrence of errors and biases that might would've altered the reliability of the research. During these time frames, all inappropriate content or behavior which may cause physical or psychological harm to younger users was identified and categorized in a designated coding instrument (Fig. 1) which was synthesized through the respective codes of conduct of the applications in combination with the different types of harassment described in Blackwell et al. [32].

We employed a categorization instrument consisting of three levels (Fig. 1), which is derived from Blackwell et al. [32] for its overarching aspects. In this study we have further specified and developed a categorization instrument which distinguishes each type of harassing/inappropriate content or behavior into verbal, physical, or environmental nature. This explains the extent to which victims encounter harassing or inappropriate activity in the metaverse while each of the three dimensions addresses characteristics unique to the metaverse that might affect each user differently. The second level refers to subcategories that further specify the reason why those situations can be seen as harassing or child inappropriate. It distinguishes between content or behavior that is insulting/offensive, sexual, violent, discriminating, self-harm promoting, or containing other child-inappropriate content or behavior (e.g., gambling, substance abuse, promotion of dangerous activities). The third level of the categorization instrument is an even more detailed specification of the previously named categories in level two. Occurrences on this level are henceforth mentioned as instances. This multi-level categorization instrument enables a nuanced understanding of the diverse forms of inappropriate content or behavior in the chosen metaverse spaces. Past research that indicates problems regarding the clear definition of what exact behavior and content can be seen as harassing and

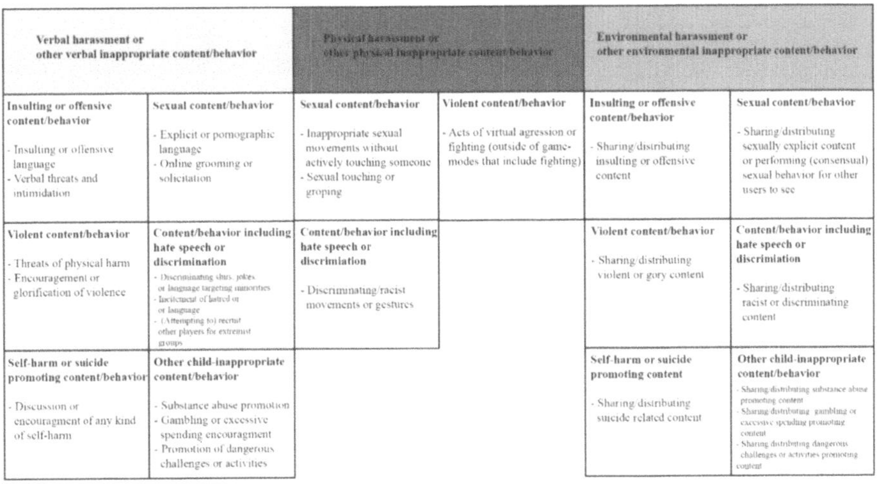

Verbal harassment or other verbal inappropriate content/behavior		Physical harassment or other physical inappropriate content/behavior		Environmental harassment or other environmental inappropriate content/behavior	
Insulting or offensive content/behavior - Insulting or offensive language - Verbal threats and intimidation	**Sexual content/behavior** - Explicit or pornographic language - Online grooming or solicitation	**Sexual content/behavior** - Inappropriate sexual movements without actively touching someone - Sexual touching or groping	**Violent content/behavior** - Acts of virtual aggression or fighting (outside of game-modes that include fighting)	**Insulting or offensive content/behavior** - Sharing/distributing insulting or offensive content	**Sexual content/behavior** - Sharing/distributing sexually explicit content or performing (consensual) sexual behavior for other users to see
Violent content/behavior - Threats of physical harm - Encouragement or glorification of violence	**Content/behavior including hate speech or discrimination** - Discriminating slurs, jokes or language targeting minorities - Incitement of hatred in or language - (Attempting to) recruit other players for extremist groups	**Content/behavior including hate speech or discrimination** - Discriminating/racist movements or gestures		**Violent content/behavior** - Sharing/distributing violent or gory content	**Content/behavior including hate speech or discrimination** - Sharing/distributing racist or discriminating content
Self-harm or suicide promoting content/behavior - Discussion or encouragement of any kind of self-harm	**Other child-inappropriate content/behavior** - Substance abuse promotion - Gambling or excessive spending encouragement - Promotion of dangerous challenges or activities			**Self-harm or suicide promoting content** - Sharing/distributing suicide-related content	**Other child-inappropriate content/behavior** - Sharing/distributing substance abuse promoting content - Sharing/distributing gambling or excessive spending promoting content - Sharing/distributing dangerous challenges or activities promoting content

Fig. 1. Structure of categorization instrument

what not [46, 47]. In order to create a basis for how online harassment is identified this paper orientates itself to the inclusive definition described by Blackwell et al. [32] which says, "Online harassment refers to a broad spectrum of abusive behaviors enabled by technology platforms". As examples, this and various other papers mentioned insults, the publication of personal information, shaming, violent threats or behavior, unwanted sexual conversations or actions, and more generally all types of offensive behavior and content that might make other users uncomfortable [32, 46, 47]. While identifying all different types of harassment we also included the presence of pornographic content, racist/discriminating behavior, or promotion of dangerous or addictive activities such as substance abuse, self-harm, or gambling [48, 49]. The overarching category of physical harassment does exclude certain sub-categories that only apply to the other two overarching categories. This was due their infeasibility to occur as a form of physical harassment. Now that we have elaborated upon our approach and categorization instrument we can present findings.

4 Results

4.1 Preliminary Interview Results

We first conducted semi-structured expert interviews to develop first conceptual ideas and a preliminary framework for the classification of the kind of risks children and adolescents could encounter in the metaverse.. Combined, the expertise of the questioned experts includes the areas of medicine, psychology, and VR-technology-related research. Table 2 displays the overview of interviewees for the exploratory interviews.

Nevertheless, certain trends or probabilities of threats still exist in their opinion. Multiple experts indicated that it is difficult to generalize potential risks since the vulnerability of a child to the dangers of the metaverse is based on an interplay of numerous subjective factors. These are for example their relationship with friends or family,

Table 2. Interviewee selection

Interviewee	Profession
A	Medical researcher in the field of addiction, genetics, and neurology
B	Lecturer with a focus on medicine and life sciences
C	Researcher with a focus on the psychiatric evaluation of VR
D	Lecturer in health psychology and technology and researcher in VR
E	Assistant professor in medicine, life sciences, and social sciences
F	Lecturer in (cyber) psychology, (e)health and digital wellbeing

predisposition for psychological problems, or their individual level of experience and self-esteem. Interviewee A stated that the research regarding the metaverse is at risk of becoming an imbalanced field of study in which researchers focus too much on its potential advantages while neglecting the dangers. Addiction is a serious concern, especially for children and adolescents. "The immersed feeling of presence in these virtual worlds is especially hard for children of a younger age to distinguish from reality. The risk of confusing real life with virtual reality becomes even stronger because our optical sense is the easiest to trick" according to interviewee A. The creation of idealized avatars can foster this occurrence even more. Since the metaverse will provide children and adolescents with an additional representation of themselves which they could prefer over their own body this could be a risk for adolescents, especially in the uncertain time of puberty. Follow-up effects that were mentioned in connection to the induced dissatisfaction with their own body were reduced self-esteem, body dysmorphia, or anorexia. Other addiction-induced effects that the experts indicated were depression, anxiety, loneliness, reduced social skills, reduced academic performance, and restlessness. Additionally, a decrease in attention skills was hypothesized by interviewee B as he indicated that "getting used to the high level of stimuli that are usually occurring in the metaverse could make it harder for children to focus on low-stimulating activities such as reading a book or following a lecture".

Regarding potential harmful intentions of other users, experts stated that due to the higher level of immersion, children might encounter traumatic incidents that could create traumas that burden them for their whole life. Interviewee E, whose research domain includes VR-induced interventions for criminals, stated that the risk of grooming or recruitment through extremist organizations might be increasingly dangerous for adolescents in the metaverse. While adolescents in puberty tend to distance themselves from their parents to search for new peer groups, it is easy for such predators or groups to provide them with a false sense of belonging and masking their true nature and intentions through personally adjusted environments and avatars. Interviewee E indicated that "terrorist groups have the possibility to create a better picture of themselves by exactly tailoring their appearance in a way that is appealing to their targets which increases the chances of recruitment compared to how it would work offline".

Also physical risks, such as cybersickness and physical injuries, were mentioned. Two experts indicated that the risk for physical injuries is significantly higher due to the impulsive and careless behavior which is often typical for children. Interviewee A mentioned that around 30% of the human's neurological connections restructure themselves during puberty and adapt to the current environmental conditions and circumstances. This might indicate that through extensive usage of the metaverse during these years, adolescents might develop characteristics specifically adjusted to the metaverse. This could range from deficits in facial recognition and social capabilities up to the disturbed satisfaction of human needs such as the search and interaction with potential sexual partners. Furthermore, interviewee A defined the time in which the neurological restructuring takes place as an "incredibly vulnerable phase for the brain in which traumas, that humans would normally process easily, could have immensely damaging effects". How should society address these kind of threats facing our youth on the metaverse? Interviewee F stated, "You wouldn't leave your child unattended in the middle of New York City. Why then in the metaverse?". Strict time restrictions and creating a balance with other activities (e.g., sports) were suggested. Strong governance systems are necessary that could actively prevent malicious individuals to harm children. Furthermore, the implementation of standardized and non-perfectionistic avatars was suggested to prevent children from preferring their virtual selves over their real bodies, while also the creation of specific child-appropriate worlds containing certain policies to protect them.

4.2 Systematic Observations in a Metaverse-Like Virtual Social Space

The following section provides an overview of situations identified during the experiments in which child-inappropriate content or behavior occurred based on systematic observation with our categorization instrument as introduced in Fig. 1.

VRChat

A total number of 19 situations were identified in which inappropriate content or behavior took place during the three hours of systematic observation in VRChat. All of these situations fell under one of the distinctions of harassment derived from Blackwell et al. [32] while simultaneously violating the community guidelines described on the application website[1]. Verbal harassment was quite a common practice in VRChat being often of sexual nature. It originated from sexually motivated conversations or actions among the avatars. This includes for example situations in which avatars engaged in sexual conversation or actions that subsequently provoked other users to use profane language towards them due to their own dissatisfaction with the situation. Other forms of verbal harassment included racist as well as physically threatening components. For the latter, the common theme was threatening other users in terms of revealing their real physical address and ultimately stating intentions to cause them physical harm. In some situations, this behavior is even channeled into an incitement for self-harm or even suicide. In addition to racist slurs and insults, one specific situation was defined by a user glorifying the actions performed by Adolf Hitler and the Third Reich while simultaneously trying to convince other avatars to follow his actions such as copying his national socialistic avatar appearance. When it comes to the promotion of substance abuse it is worth mentioning that the identified situations all included multiple users instead of one lone

---

[1] hello.vrchat.com/community-guidelines.

perpetrator. Those included group conversations of users that either already consumed alcohol and tried to encourage others to do the same or the glorification of drugs such as LSD. Regarding physical harassment between avatars, one situation occurred in which one user pretended to physically assault another during an argument after which the victim left the game. Furthermore, multiple situations of clearly nonconsensual groping and sexual touching were identified where the unwillingness of the victim was obvious due to their complaints of the situation via voice chat. One clear example of this was a user that described himself as an adult male and repeatedly engaged in sexual talk and groping towards another user who claimed to be an underage girl. This situation can be seen as an attempt for online grooming. Even after the age of the girl was known the adult man did not stop his attempts of trying to persuade her. When it comes to environmental harassment, various situations were categorized in which avatars engaged in sexual interaction, conversations, and drawings for the whole space to witness. This comprised simple touching of explicit body parts as well as an "virtual orgy" including multiple characters performing sexual positions. Besides that, another environmental harassment that took place refers to the previously mentioned situation of a user who changed the appearance of his avatar to one that incorporates swastikas for everyone to see.

RecRoom

RecRoom contained relatively fewer inappropriate situations during the 3-h of play. Six indications of harassment were found during the observation. No occurrences of physical harassment were detected. However, this does not necessarily mean that it does not exist in this space, but seem less common for this platform. All types of verbal harassment took place in the context of a game. These consisted of verbal insults due to defeat, teasing the opponent during a fight through profane language, and threats of physical violence. The latter occurred during a game mode after the opponent already repeatedly mentioned swear words, some of them including racist expression. Furthermore, one situation took place in which one user vocalized a homophobic statement in the context of a group conversation regarding gender fluidity. Given the situation that one game mode provided the possibility to interact with virtual alcohol, multiple users began to drink the artificial beverages, act drunk, and encouraged others to participate. This represents a case of environmental child inappropriate behavior due to the clear glorification and incitement of alcohol consumption. Another situation worth mentioning is the broadcasting of an explicit song by a user for all other members of the space to hear. While in itself the unwelcome broadcasting of music can be already seen as a form of harassment, this song included various insults, racist slurs, glorification of illegal substances, and overall profane language.

5 Discussion

The classification of child inappropriate behavior and risks of this paper provides researchers with a classification method which can be utilized for future content analyses of any metaverse space (Fig. 1). With the focus laying on the categorization of child-inappropriate content or behavior, this framework consists of an elaborate catalog to differentiate between occurrences of various natures as well as to make a statement regarding their frequency during the duration of the experiment. In a practical sense, this paper could facilitate other researchers, policymakers, and developers to build upon this categorization framework for developing governance systems that actively restrain these identified risks from occurring. However, this classification method should be further empirically tested to ensure its validity and potentially extent it with more categories. We encourage studies on a larger scale to acquire more quantitative data as well. This contributes to fostering a safe digital environment that offers enriching experiences while mitigating potential harm.

The most prominent limitation of this study was the amount of time spent inside each of the metaverse spaces. While the categorization instrument is easily applicable to long-term studies, the three hours spent in each space cannot be seen as representative enough to make any quantitative claims regarding the average frequency of inappropriate content or behavior identified or the potential difference between the two spaces. Due to a lack of time and the nature of the thesis it was not possible to further increase the time spent in each of the spaces, as well as the amount of spaces that were examined. But at this we did not strive for any quantitative numbers but merely wanted to increase theoretical understanding of the possibilities of harm. Significant differences might occur if the study duration as well as the tested metaverse spaces were increased. Furthermore, inside both applications, numerous subspaces exist that might be defined by different user segments and behavioral dynamics. The goal of this study was to provide an exemplary overview of what inappropriate situations children and adolescents can possibly encounter during their time in the metaverse. Another factor that became apparent during the research was that the application's current self-governance systems are mainly user-centric and depending on user-reporting. The real-time voice chat moderation included in RecRoom is the only identifiable method which does not rely on users' participation. Nevertheless, it did not fully restrict inappropriate language from occurring. In regard to a functioning age limitation, both applications propose a minimum age of 13 years old but fail to implement a form of identity verification in order to enforce this limitation and instead rely on self-reporting by users. The available time and scope of the study also acted as a limitation by inhibiting the comparison of various governance systems implemented in other metaverse spaces. This might have given rise to a more detailed evaluation and recommendations for improvement. While the scope of the conducted research focused on what inappropriate content or behavior can be identified in the metaverse that can directly impact the user's psychological and physiological health, it excluded the numerous amounts of privacy concerns this new technology brings with it [1, 5, 7, 10, 50, 51]. Furthermore, the chosen data collection method could of course not detect health related cases such as motion sickness and physical harm during the activity of other users. Finally, even though the chosen platforms of RecRoom and VR Chat offer quite some functionality that will be apparent in the future metaverse, these systems are falling short in delivering a digital twin experience and lack augmented reality elements as expected in future versions of such applications.

6 Conclusion

The findings of this research showed the existence of numerous types of child-inappropriate content or behavior. These are for example of insulting, sexual, violent, or discriminating nature while ranging from verbal over physical up to environmental types of harassment. Currently, online applications that are the candidates for metaverse do not provide a safe environment for children and adolescents to take part in. The prevalence of child-inappropriate content or behavior is evident and must not be ignored by parents, developers, or policymakers. Psychological or physiological harms might become the result of minors entering virtual worlds potentially leading to long-term consequences. Compared with the closely related concepts of social media and online gaming a clear tendency is observable that the metaverse recreates similar issues that were already examined in those prior online spaces. Additionally, aggravated by the heightened factors of immersion, presence, and embodiment more severe follow-up effects of harassing and inappropriate content or behavior are expected. By experiencing traumas that could profoundly impact their entire life, minors are at risk of suffering from altered brain development and lasting emotional and physical distress. Moreover, our study highlights the shortcomings of the examined governance systems to provide sufficient enforceability of their own code of conduct[2]. The lack of active moderation, coupled with ineffective enforcement of rules and a lack of immediate punishment for perpetrators, leaves children and adolescents vulnerable to harm. It is imperative for platform owners to take responsibility and implement governance measures that holds malicious users accountable and actively safeguards younger users.

After the potential effects of cyberbullying are studied extensively in literature it becomes clear how impactful and dangerous the effects of harassment will be for children and adolescents in the metaverse. Known outcomes for their psychological and physiological health include stress, anxiety, depression, loneliness, substance abuse, emotional distress, anger, and most concerning a positive relationship with suicidal ideation among victims [53, 54]. The latter might be even increased due to the identified encouragement and promotion of self-harm or suicide that occurred in the social space of "VRChat". The heightened feeling of immersion and presence induced by the technological characteristics often attributed to the metaverse have the potential to further amplify the mentioned effects for younger users. As mentioned by Goltz [19] and Lee et al. [10] the "blurring of borders" between VR and real life, which especially children are at risk of, could either intensify negative effects while simultaneously creating the risk of adapting certain malicious behaviors and projecting them into real life. This aligns with the research of Lavoie et al. [9], which states how immersion in a virtual space can intensify negative emotions and even in the case of younger users. Since the concept of the metaverse is anticipating to provide users with an alternative reality which will co-exist with real life in the form of a virtual mirror world, the extent of cyberbullying might even exceed the current impacts created by traditional social media or gaming. This is due to its potential of becoming a highly immersive space in which children and adolescents spend a majority of their time [5, 10].

---

[2] recroom.com/code-of-conduct.

Furthermore, the possibility of sexual harassment became evident during the systematic observations in "VRChat". Multiple situations included the use of highly sexualized language, explicit actions performed amongst other users, or the attempt of older users to groom minors. These are all situations that are imperative to be seen as child-inappropriate due to their potential of causing long-term emotional distress, trauma, and problematic sexual behavior. In this context, the possibility exists that adolescents become used to certain characteristics, behavioral dynamics, or preferences that are prevalent or unique to the metaverse. These might include substance abuse, aggressive behavior, racist behavior, or the development of sexual preferences linked to virtual worlds or characters. These are all aspects that were possible to encounter during the carried-out method of systematic observation based on our classification instrument.

These findings may facilitate collaborations between researchers, policymakers, and developers of this technology to prepare for participation of youth in the metaverse. We should ensure the well-being of users and the creation of innovative solutions that mitigate the identified risks and challenges, fostering a secure digital environment that is safe for all user segments. We encourage other researchers to further study this topic and help to shed light on the serious risks of digital metaverse spaces. Together with practitioners it is important to develop governance measures that help to protect younger generations when they are entering the future metaverse.

References

1. Allam, Z., Sharifi, A., Bibri, S.E., Jones, D.S., Krogstie, J.: The metaverse as a virtual form of smart cities: opportunities and challenges for environmental, economic, and social sustainability in urban futures. Smart Cities **5**, 771–801 (2022)
2. Abbate, S., Centobelli, P., Cerchione, R., Oropallo, E., Riccio, E.: A first bibliometric literature review on Metaverse. In: 2022 IEEE Technology and Engineering Management Conference (TEMSCON EUROPE), pp. 254–260 (2022)
3. Stephenson, N.: Snow Crash. Bantam Books, New York (1992)
4. Horizon Worlds Virtual Reality Worlds and Communities. https://www.meta.com/horizon-worlds/
5. Dwivedi, Y.K., et al.: Metaverse beyond the hype: multidisciplinary perspectives on emerging challenges, opportunities, and agenda for research, practice and policy. Int. J. Inf. Manage. **66**, 102542 (2022)
6. Damar, M.: Metaverse shape of your life for future: a bibliometric snapshot. J. Metaverse. **1**, 1–78 (2021)
7. Bibri, S.E., Allam, Z.: The metaverse as a virtual form of data-driven smart urbanism: on post-pandemic governance through the prism of the logic of surveillance capitalism. Smart Cities. **5** (2022)
8. Kaimara, P., Oikonomou, A., Deliyannis, I.: Could virtual reality applications pose real risks to children and adolescents? A systematic review of ethical issues and concerns. Virtual Real. **26**, 697–735 (2022)
9. Lavoie, R., Main, K., King, C., King, D.: Virtual experience, real consequences: the potential negative emotional consequences of virtual reality gameplay. Virtual Real. **25**, 69–81 (2021)
10. Lee, L.-H.: All one needs to know about metaverse: a complete survey on technological singularity, virtual ecosystem, and research agenda. J. Latex Class Files **14** (2021)

11. Muslihati Hotifah, Y., Hidayat, W.N., Purwanta, E., Valdez, A.V., Ilmi, A.M., Saputra, N.M.A.: Predicting the mental health quality of adolescents with intensive exposure to metaverse and its counseling recommendations in a multicultural context. Cakrawala Pendidikan. **42**, 38–52 (2023)
12. Usmani, S.S., Sharath, M., Mehendale, M.: Future of mental health in the metaverse (2022)
13. Bailey, J.O., Bailenson, J.N.: Considering virtual reality in children's lives. J. Children Media **11**(1), 107–113 (2017)
14. Park, S.M., Kim, Y.G.: A metaverse: taxonomy, components, applications, and open challenges. IEEE Access. **10**, 4209–4251 (2022)
15. Benrimoh, D., Chheda, F.D., Margolese, H.C.: The best predictor of the future—the metaverse, mental health, and lessons learned from current technologies. JMIR Ment. Health, **9** (2022)
16. Cairns, P., Cox, A., Nordin, A.I.: Immersion in Digital Games: Review of Gaming Experience Research. In: Angelides, M.C., Agius, H. (eds.) Handbook of Digital Games, pp. 337–361. John Wiley & Sons Inc, Hoboken, NJ, USA (2014)
17. Han, D.I.D., Bergs, Y., Moorhouse, N.: Virtual reality consumer experience escapes: preparing for the metaverse. Virtual Real. **26**, 1443–1458 (2022)
18. Shin, D.: Empathy and embodied experience in virtual environment: to what extent can virtual reality stimulate empathy and embodied experience? Comput. Human Behav. **78**, 64–73 (2018)
19. Goltz, N.: ESRB warning: use of virtual worlds by children may result in addiction and blurring of borders. Pittsburgh J. Technol. Law Policy. **11** (2011)
20. Anderson, C.A., et al.: Supplemental material for violent video game effects on aggression, empathy, and prosocial behavior in eastern and western countries: a meta-analytic review. Psychol. Bull. (2010)
21. Griffiths, M.D., Davies, M.N.O., Chappell, D.: Online computer gaming: a comparison of adolescent and adult gamers. J. Adolesc. **27**, 87–96 (2004)
22. Witmer, B.G., Singer, M.J.: Measuring presence in virtual environments: a presence questionnaire. Presence: Teleoperators Virtual Environ. **7**, 225–240 (1998)
23. Turel, O., Romashkin, A., Morrison, K.M.: A model linking video gaming, sleep quality, sweet drinks consumption and obesity among children and youth. Clin. Obes. **7**, 191–198 (2017)
24. Gottschalk, F.: Impacts of technology use on children: Exploring literature on the brain, cognition and well-being. OECD Education Working Papers (2019)
25. Kenney, E.L., Gortmaker, S.L.: United States adolescents' television, computer, videogame, smartphone, and tablet use: associations with sugary drinks, sleep, physical activity, and obesity. J. Pediatr. **182**, 144–149 (2016)
26. Dey, N., Babo, R., Ashour, A.S., Bhatnagar, V., Bouhlel, M.S.: Social Networks Science: Design, Implementation, Security, and Challenges: From Social Networks Analysis to Social Networks Intelligence. Springer, Cham (2018)
27. Holsapple, C.W., Wu, J.: User acceptance of virtual worlds: the hedonic framework. DATA BASE Adv. Inf. Syst. **38**, 86–89 (2007)
28. Fox, J., Bailenson, J.N., Tricase, L.: The embodiment of sexualized virtual selves: the Proteus effect and experiences of self-objectification via avatars. Comput. Human Behav. **29**, 930–938 (2013)
29. Paul, I., Mohanty, S., Sengupta, R.: The role of social virtual world in increasing psychological resilience during the on-going COVID-19 pandemic. Comput. Human Behav. **127**, 107036 (2022)
30. Yee, N., Bailenson, J.N., Ducheneaut, N.: The proteus effect: implications of transformed digital self-representation on online and offline behavior. Commun. Res. **36**, 285–312 (2009)
31. Teng, C.-I.: Customization, immersion satisfaction, and online gamer loyalty. Comput. Human Behav. **26**, 1547–1554 (2010)

32. Blackwell, L., Ellison, N., Elliott-Deflo, N., Schwartz, R.: Harassment in social virtual reality: challenges for platform governance. Proc. ACM Human Comput. Interact. **3**, 1–25 (2019)
33. Chen, G.M., Pain, P., Chen, V.Y., Mekelburg, M., Springer, N., Troger, F.: 'You really have to have a thick skin': a cross-cultural perspective on how online harassment influences female journalists. Journalism **21**, 877–895 (2020)
34. O'Keeffe, G.S., et al.: The impact of social media on children, adolescents, and families. Pediatrics **127**, 800–804 (2011)
35. Shriram, K., Schwartz, R.: All are welcome: using VR ethnography to explore harassment behavior in immersive social virtual reality. In: 2017 IEEE Virtual Reality (VR), pp. 225–226. IEEE, Los Angeles, CA, USA (2017)
36. Jonsson, L.S., Fredlund, C., Priebe, G., Wadsby, M., Svedin, C.G.: Online sexual abuse of adolescents by a perpetrator met online: a cross-sectional study. Child Adolesc. Psychiatry Ment. Health **13**, 1–10 (2019)
37. Lawson, E.C.: New research shows Metaverse is not safe for kids (2021). https://counterhate.com/blog/new-research-shows-metaverse-is-not-safe-for-kids/
38. Kartit, D., Howcroft, E., Howcroft, E.: Interpol says metaverse opens up new world of cybercrime. Reuters (2022)
39. Koehler, D., Fiebig, V., Jugl, I.: From gaming to hating: extreme-right ideological indoctrination and mobilization for violence of children on online gaming platforms. Polit. Psychol. **44**, 419–434 (2023)
40. Nichols, S., Patel, H.: Health and safety implications of virtual reality: a review of empirical evidence. Appl. Ergon. **33**, 251–271 (2002)
41. Regan, C.: An investigation into nausea and other side-effects of head-coupled immersive virtual reality. Virtual Real. **1**, 17–31 (1995)
42. Cao, S., Nandakumar, K., Babu, R., Thompson, B.: Game play in virtual reality driving simulation involving head-mounted display and comparison to desktop display. Virtual Real. **24**, 503–513 (2020)
43. LaViola, J.J.: A discussion of cybersickness in virtual environments. ACM SIGCHI Bull. **32**, 47–56 (2000)
44. Cloete, R., Norval, C., Singh, J.: A call for auditable virtual, augmented and mixed reality. In: Presented at the November (2020)
45. Popper, K.: The Logic of Scientific Discovery. Taylor & Francis, Milton Park (2005)
46. Duggan, M.: Online Harassment 2017 (2017). https://www.pewresearch.org/internet/2017/07/11/online-harassment-2017/
47. Wolak, J., Mitchell, K.J., Finkelhor, D.: Does online harassment constitute bullying? An exploration of online harassment by known peers and online-only contacts. J. Adolesc. Health. **41**, S51–S58 (2007)
48. Strasburger, V.C., Jordan, A.B., Donnerstein, E.: Children, adolescents, and the media: health effects. Pediatr. Clin. North Am. **59**, 533–587 (2012)
49. Weimann, G., Masri, N.: Research note: spreading hate on TikTok. Studies Conflict Terrorism **46**, 752–765 (2023)
50. Falchuk, B., Loeb, S., Neff, R.: The social metaverse: battle for privacy. IEEE Technol. Soc. Mag. **37**, 52–61 (2018)
51. Fernandez, C.B., Hui, P.: Life, the metaverse and everything: an overview of privacy, ethics, and governance in metaverse. In: Presented at the March (2022)

Navigating Privacy Regulations: Administrative Burden of Digital Self-Services for Vulnerable Citizens

Ida Heggertveit(✉)

University of Agder, Kristiansand, Norway
ida.heggertveit@uia.no

Abstract. Balancing privacy regulations and usability is a challenge in the delivery of government welfare services through digital self-services. This study explores the experiences of vulnerable citizens with digital self-services in government welfare services, focusing on the administrative burdens of privacy regulation vulnerable citizens experience when trying to access and use the benefits for which they are eligible. Through interviews, focus groups and observations, three key costs - compliance, psychological, and learning – are examined. The findings shed light on how privacy regulation worsens the burden and usability of self-services and reproduce socioeconomic inequalities for vulnerable citizens by (1) limiting access to support, (2) putting them at risk of identity theft and fraud, and (3) demanding excessive documentation with limited opportunities to resubmit.

Keywords: Administrative Burden · Vulnerable Citizens · Digital Self-Service · Usability · Privacy Regulation

1 Introduction

Digital government agencies strive to provide citizens with user-friendly, democratic, transparent, and effective digital self-services that improve service delivery and save costs for the agency [1]. This changes the public encounter because self-services allow citizens to take an active role in managing administrative tasks themselves, which includes accessing information, submitting applications, and engaging with the government online trough different digital channels [2–5]. This has inspired governments worldwide to implement digital self-service options.

However, digital government implementations can cause poor usability because they fail to consider citizens needs and life-situations [6–11] *Usability*, in this context, refers to the extent to which self-services can be successfully, efficiently, and satisfactorily utilized by citizens [12]. Additionally, the successful implementation of digital self-services relies on a secure information structure that respects privacy regulations [13], as these services can create new administrative challenges for citizens who must understand the risks and responsibilities involved, especially regarding privacy regulations [14–16] *Privacy regulations* govern the control, use, and distribution of citizens' digital information, but there is a research gap in understanding the specific challenges citizens face in balancing usability and privacy regulations, particularly in the context of self-services.

When designing self-service programs, government agencies tend to overlook the circumstances of vulnerable citizens in difficult life situations who rely on social safety net programs and other welfare benefits. These agencies typically design self-services for the mainstream [17], leaving vulnerable citizens to face cognitive challenges and higher administrative costs. As a result, vulnerable citizens can perceive self-services as an onerous policy implementation [18, 19] and unequal distribution of administrative burdens reinforces social inequalities and leads to negative government-citizen interactions for vulnerable citizens [20–22].

The study highlights a gap between citizens' needs and the self-services designed to fulfill those needs [23, 24]. By investigating administrative burden, it sheds light on the challenging experiences citizens have when dealing with welfare services, particularly in terms of learning, compliance, and psychological aspects [18]. These costs are rooted in policy implementations such as organizational structure, document requirements, and a "digital-first" strategy that aims for increased use of digital technology in all communication with citizens [25], all of which are subject to privacy regulations designed to protect vulnerable citizens' privacy. Therefore, guided by this research question: *What challenges of privacy regulations related to digital self-services do vulnerable citizens experience?* This study examines vulnerable citizens experiences of privacy regulations and how they relate to usability and their access to welfare services through self-services.

The empirical cases of this study are two Norwegian welfare services: (1) the social safety net benefit *financial support* for citizens who have no other financial resources and (2) the welfare benefit *physical aid and accessibility* for disabled and/or chronically ill citizens seeking to enhance their quality of life. These services are offered digitally through self-service channels. The study employs various research techniques, including observation, interviews, and focus groups, to gather insights from citizens who utilize these services. The paper begins with an introduction, followed by a research background with the theoretical framework that outlines the analytical approach. Next, the methodology section provides details on the setting of the study, data collection methods and analytical approach. The results and discussion section presents examples from the findings that highlights how privacy regulations impose administrative burdens on vulnerable citizens when using digital self-services seen in the light of previous research. Finally, it presents the conclusion, limitations and recommendations for future research directions.

2 Research Background

This section introduces the vulnerable citizens who use digital self-services in this research, as well as an overview of the concept of administrative burdens they experience. The literature presented in this section serves as background information for this study.

2.1 Vulnerable Citizens

Government agencies prioritize digital self-service as the primary channel for citizens to interact with them because it offers easier access to welfare services [5]. It is important to acknowledge the experiences of vulnerable citizens when analyzing these self-services as their unique complexities are often overlooked [16, 26], and they often encounter higher levels of complexity and challenges with government programs [22]. In this study, *vulnerability* refers to social dependence in difficult life-situations [27]. In this study, vulnerable citizens often experience scarcity and depend on social safety net programs and other welfare benefits because they are: (1) unemployed citizens seeking financial support and (2) parent caregivers of disabled and/or chronically ill children seeking physical aid. *Scarcity* can mean shortage of physical resources (e.g., money, network), or shortage of cognitive recourses (e.g., attention, executive control). The cognitive part is particularly problematic for vulnerable citizens who need to apply on their own through digital self-services because it affects their performance in completing administrative tasks [19].

Scarcity can significantly impact the experiences of vulnerable citizens, leading to increased stress, reduced executive functioning, and difficulties navigating interactions with the government. Managing self-services becomes more challenging for citizens in vulnerable circumstances, such as those seeking financial support, who are often facing challenging life situations already [28–30]. Parent caregivers of disabled and/or chronically ill children are also considered vulnerable due to their increased risk of experiencing prolonged burden and symptoms of burnout [31]. These parents face numerous challenges and overwhelming responsibilities in their lives [32], impacting their health-related quality of life, social interactions, and financial responsibilities. As they require support to manage daily expenses and access essential services, they are often the focus of government programs [33, 34]. Both groups can experience physical and psychological fatigue which can also hinder their ability to navigate digital self-services effectively [16] leading them to rely on *gatekeepers* (i.e., frontline workers who control access to services, benefits and knowledge) who they hope will help reduce the burden [35].

2.2 Administrative Burden

The struggles vulnerable citizens experience with self-services can be understood through the concept of administrative burden [18]. In essence, administrative burden relates to a citizen's perception the process of carrying out a policy or following certain administrative procedures is burdensome or demanding [34]. Initially this framework was used to analyze employees' experiences with policy tasks in the public sector, but Herd and Moynihan later expanded this concept to include the various costs that *citizens* might encounter when engaging with government agencies. More recently, scholars in the field of information systems have broadened the scope of this concept to include the domain of digital government [16, 30, 36, 37].

Administrative burdens can create structural inequalities to public rights and services. These structural inequalities can be conceptualized as three different types of *costs* citizens face when engaging with government entities. *Compliance costs* related to information and access, *psychological costs* associated with stress and loss of autonomy, and *learning costs* requiring significant time and effort investment [18].

This typology has several advantages. It makes the costs associated with administrative burden clearer and more identifiable, creating a distinct concept that is easier to comprehend. It also enables the illustration of specific types of burdens and demonstrates how the costs encountered by citizens are related to government issues and political decisions. Additionally, the typology provides a valuable diagnostic tool that conceptualizes specific types of burdens and makes it easier to identify when and where they arise, as well as their impact on the lives of citizens [38]. Examining the privacy regulations faced by vulnerable citizens when utilizing self-services through the theoretical lens of administrative burdens is especially relevant for those applying for financial support and physical aid. The data and method section will describe how the experiences of vulnerable citizens regarding burdens are captured.

3 Empirical Context

This research centers on Norway's welfare sector, which is well known globally for its advanced digital public welfare services and consistently high rankings on the UN's Human Development Reports. The core of Norway's welfare system is The Norwegian Labour and Welfare Administration (NAV), responsible for delivering and administering all labor and social security benefits. NAV provides financial support to citizens in economic need and accessibility technology to citizens with physical/cognitive disabilities, referred to as Physical Aid. These two services were selected for this study because they are good examples of how local government services can complicate the usability of services when they are presented in a one-stop shop with a channel management system built for central government services.

NAV has implemented a unified service model, known as the one-stop shop for NAV Offices in each municipality, which consolidates local and central government services into a national system accessible through a single point of entry for citizens [39]. While the NAV office is a shared responsibility between the central and local government, each entity maintains autonomy over its respective services. The central government entity focuses on employment-related welfare services, benefits, and pensions, whereas the local government entity manages social services, including financial assistance, with dedicated budgets. The NAV Physical Aid Center in each county, operated by the central government, is responsible for providing physical aid to all citizens. Meanwhile, the local government handles communication with the citizens, facilitates the application process on their behalf, and is responsible for educating citizens on how physical aid functions.

NAV is often considered a 'best case' within public sector digitalization [40] where citizens are redirected from expensive face-to-face interactions towards cost-reductive digital channels, creating a *faceless interaction* with NAV employees through chat, chatbot, web, and mail [41]. NAVs channel strategy aligns with the Norwegian government's *digital first* strategy, where the goal is to streamline access to welfare services, encouraging citizens to use self-services whenever available [25]. NAV, responsible for managing a significant portion of the national budget, possesses one of the most extensive collections of personal and sensitive data about Norwegian citizens, and this study explores the administrative challenges related to these personal and sensitive data within self-service applications for vulnerable citizens applying digitally for local government welfare services.

3.1 Data Collection

Table 1 provides an overview of the data collection, involving 14 instances of observations with unemployed citizens applying for financial support at the local NAV office. I closely monitored their utilization of the digital self-service platform and conducted interviews to explore deeper into their experiences and to gain insights into the administrative burdens they faced during the application process.

To gain a further in-depth understanding of how citizens experience the application [42] I participated in a research group where we conducted three focus groups with citizens who had knowledge of digital self-services applications and firsthand experience navigating similar service procedures within NAV [43]. The first focus group involved unemployed immigrant women participating in a local government program aimed at facilitating their transition into employment or education. Subsequently, we conducted two additional focus groups, both groups with parents of chronically ill/disabled children. Being part of the research group, our primary objective of these focus groups was to get citizens' insights and experiences regarding the utilization of digital self-services for welfare. We therefore structured the discussions around findings we earlier gained from interviews and observation. The participants were guided through a series of key questions addressing various aspects of the application process, enabling in-depth exploration and reflection on their experiences. Subsequently, the groups engaged in discussions, reflected on the presented data, and responded to our follow-up inquiries.

Table 1. Empirical Data: An Overview

	Respondent	Code	Period
Observations & Interviews	14 Citizens applying for financial support	OB1-OB14	Autumn 2023
Focus Groups	5 Unemployed Immigrant women 3 Parents of chronically ill/disabled children 3 Parents of chronically ill/disabled children	OB15-OB19 OB20-OB23 OB24-OB26	Autumn 2023 Spring 2024 Spring 2024

3.2 Data Analysis

I started the analysis of the data collection by extracting direct quotes about privacy concerns from the data, followed by a six-phase thematic analysis approach [44], utilizing MAXQDA software. First, I familiarized myself with the data by spending hours transcribing, listening to interviews, focus groups, and reviewing observation notes. Second, I used the software to generate codes for identifying privacy-related burdens experienced by citizens. Third, I conducted a search for keywords, such as "assistance," "access," and "bankID" (personal digital identification device) which formed the basis for themes. Fourth, I examined these themes and identified additional themes related to privacy regulations in digital self-services. I then defined new themes, such as "information exchange" and "reuse documentation". Finally, I documented the findings. This method facilitated a comprehensive informative examination of the everyday social processes [45] experienced by vulnerable citizens and families, particularly in managing health and welfare services—a routine aspect of their lives.

4 Results and Discussion

Table 2 provides a short summary of the burdens and costs related to privacy regulations that were identified through observations, interviews, and focus group discussions with vulnerable citizens applying for welfare services using digital self-services.

Table 2. Summary of Findings

Type of cost	Administrative Burden of Privacy Regulation
Compliance cost	The organizational structure of a one-stop-shop and channel strategy designed for central government services, along with privacy regulations that limit the exchange of information between central and local officials, leads to high compliance costs such as limited access to support for citizens applying for financial assistance and physical aid through digital self-services. These regulations are meant to protect citizens, but they can also be a source of burden for those who rely on local government services
Psychological cost	The requirement of using digital identification devices for digital self-services exposes vulnerable citizens who depend on third-party assistance to the risk of identity theft and fraud. This not only compromises their personal security but also leaves them feeling powerless, with limited resources, and a loss of autonomy
Learning cost	Vulnerable citizens face high demands for documentation, with limited opportunities to resubmit these documents due to privacy regulations. This can be particularly challenging for citizens who may face difficulties in gathering and submitting the required documentation

One notable finding from this research is that privacy regulations present a significant burden for vulnerable citizens. Interestingly, it is not because they are overly concerned about their privacy, but rather because the burdens created by the government's efforts to protect privacy led citizens to compromise their privacy to access user-friendly self-services. This paradox arises from the fact that vulnerable citizens, who heavily rely on welfare services, are subjected to strict privacy regulations that are intended to safeguard their privacy, but which end up hindering their access to suitable support and guidance when using digital self-services. Consequently, these citizens experience additional administrative burdens, adding to their already stressful situations.

This study addresses a research gap by examining the specific factors that contribute to the increased burden faced by vulnerable citizens when using digital self-services [21]. The findings underscore the crucial role of privacy regulations in shaping vulnerable citizens' perception of the usability of these services [46], and the analysis uncovered various burdens, highlighting the delicate balance between privacy and usability in welfare services [9, 13]. Next, I will provide examples from citizens' experiences during the application process that demonstrate how administrative practices shape compliance, psychological, and learning costs.

4.1 Compliance Costs - Information Exchange

Compliance cost describes the amount of work the citizens must put into the process of applying for financial support and physical aid using the digital self-service solution [46]. The participants of this study faced challenges when it came to completing and submitting forms needed for applying for welfare and social safety net benefits, which is in line with previous research [47, 48], demonstrating that vulnerable citizens require assistance to utilize digital self-services. However, organizational structures and NAV's channel strategy result in additional compliance costs for vulnerable citizens, making it difficult to seek help from NAV. Citizens applying for central government services have access to numerous digital channels, such as personalized messaging, chat, chatbots, "share screen," and NAV call center assistance. However, due to privacy regulations, these channels are not accessible to citizens applying for local government services such as social safety net programs. As a result, they do not receive similar levels of service. This highlights previous research showing that citizens applying for means-tested services experience higher administrative costs than those using universal services [18].

The implementation of NAV's one-stop-shop and channel strategy, which excludes local government social safety net programs, has made communication between citizens and caseworkers a significant challenge. Citizens must first contact NAV's frontline call center, where frontline workers (central government officials) function as gatekeepers to caseworkers responsible for the management of financial support services [35]. Privacy regulations make it challenging for frontline workers to document information related to social services, consequently hindering their ability to communicate on behalf of citizens effectively. Verbal communication becomes necessary, resulting in frontline workers initiating phone calls to caseworkers presenting the citizens' arguments. Overall, this channel strategy creates compliance costs for citizens as they are required to repeat their story to gatekeepers hoping they present the right arguments on their behalf [35]. In cases where caseworkers need clarification on arguments, they need to call citizens within two working days, adding to citizens' compliance costs of waiting for help. While there are alternative digital channels such as chat and "share screen," they are not considered secure options for dealing with financial support or physical aid matters. Citizens who prefer these channels often express frustration as they are limited in terms of functionality and cannot be utilized for sensitive information exchanges.

Citizens applying for Physical Aid also meet challenges in applying through digital self-services. Normally citizens do not apply themselves; NAV recommends citizens get help from an authorized professional e.g., a local government healthcare worker or occupational therapist to submit the digital application on their behalf. However, after the professional has conducted the application, they have no insight into the decision made by the caseworker. This information is only possible for the citizen to access on their personal site online, and parents cannot access their children's site due to privacy regulations – leaving both the professional and the citizens in the dark.

To sum up, the organizational structure within NAV, with its division of responsibility between central and local entities within the organization, poses significant challenges for vulnerable citizens accessing digital self-services. While the duty of confidentiality ensures data security, it complicates communication between citizens and the organization. NAV's channel strategy, despite aiming to provide "digital-first" services, excludes certain programs, such as social safety net benefits from local governments, due to privacy regulations [49, 50]. The implementation of a "one-stop shop" intended to improve coordination between local and central entities delivering welfare services unfortunately creates administrative burdens for vulnerable citizens [39]. The decision to transition all communication online may have been driven by political motivations, but it clearly increases administrative burdens for vulnerable citizens by limiting direct contact with their caseworkers. Consequently, vulnerable citizens experience heightened compliance costs when seeking help with applying for social safety net programs through digital self-services.

4.2 Psychological Cost - Insufficient Help

Psychological costs describe what actions citizens feel compelled to do that leads to lack of autonomy, loss of power and the feeling of not having enough resources [18]. The psychological cost arises from feeling stressed or uncomfortable when using digital self-services without the support and guidance of a third-party. Vulnerable citizens often require assistance from others to navigate welfare services and examples from the findings show how they are put at risk of fraud and identity theft and how they lose their sense of autonomy when sharing sensitive personal information with a third-party in exchange for benefits.

Digital self-services often require citizens to manage their own applications, using personal digital identification devices. However, this study's findings show that citizens often seek assistance from various third parties, including family members, friends, and frontline workers at local NAV offices. When struggling to complete digital applications for financial support, citizens often visit their local NAV office, where frontline workers may resort to workarounds. For instance, they may take control of the computer in the self-service area and complete applications on behalf of citizens. However, ethical concerns arise regarding this practice, particularly when applications involve secure logins, and it raises questions about the level of understanding and engagement citizens have with the application process of a *self*-service when completed by others on their behalf. Additionally, citizens' privacy is exposed to other individuals in the self-serving area at the office, where there may be only one printer and copy machine, leading to personal documents frequently getting lost, exposed or forgotten.

In certain situations, citizens may entrust their personal digital identification devices to individuals within their network. For instance, children with better language skills might utilize their parents' digital identification device to apply for services on their parents' behalf. Despite privacy regulations, citizens who face language barriers or other challenges in applying for financial support often heavily rely on assistance from their children, compromising their privacy protections and placing stress on their children in the process to alleviate their administrative burdens. Privacy regulations also come into play when family and friends assisting with digital applications contact the NAV call center on behalf of the citizens. In some cases, they may even impersonate the citizens to ensure accurate assistance from frontline workers. Although frontline workers may identify impersonation and terminate the call, this does not prevent further attempts. When accessing *central* government services through NAV, citizens have the option to log onto the website and create a power of attorney, which allows their designated helper to digitally access documents and applications. However, this solution does not extend to local services such as social safety net programs, which are often the ones where citizens require assistance the most.

Parents play a crucial role as third-party assistants when helping their children apply for physical aid benefits. However, parents face limitations when they need to contact NAV to apply for assistance for their children's application, as the aid is registered under the child's name, and parents cannot access their child's information digitally. This creates frustration for parents who require written responses from NAV as they are feeling overwhelmed with information and administrative task, where having it in written can help sort the information at a later time [32].

One parent expressed frustrations, saying, *"I don't understand why NAV can provide oral answers, but not written ones to parents"* (OB25- Parent of chronical ill/disabled children) as their child's NAV page requires a personal digital device that their six-year-old child does not have. Additionally, parents are required to open a bank account in the child's name to receive benefits, even though the one paying the bills is the parent until the child becomes an adult. One parent mentioned, *"I had to open an account for my son when he was 10 months old."* (OB24 - Parent of chronical ill/disabled children).

Despite having limited insight into their child's case, parents are responsible for applying and submitting all necessary documentation. Another parent highlighted the challenges, saying, *"I only communicate with NAV in writing because we are responsible, but it's frustrating that benefits have to be discussed over the phone, and we don't receive a transcript because it concerns my child, and I don't have access. We cannot communicate in writing on behalf of our children, because they don't have a bank ID. We receive decisions addressed to (us) by post, unable to access them digitally."* (OB26 - Parent of chronical ill/disabled children). Communication with NAV is limited, as benefit discussions must take place over the phone, and parents do not receive a transcript over the phone call because of their child's privacy making them feel powerless if they need to argue/complain later for the service they applied for. Communication in writing on behalf of children is impossible because they do not have a personal digital identification device.

In summary, vulnerable citizens often rely on various third parties, including family members, friends, and central government officials, to navigate welfare services, even if it means compromising their privacy. Digital self-services require vulnerable citizens to have personal digital identification devices to access the benefits they are entitled to. This means that vulnerable citizens who depend on social safety net benefit may have to expose their personal data or hand over their personal digital identification devices to third parties in order to access these benefits, which raises privacy concerns. The complexities of identity management systems, combined with privacy regulations, present significant usability challenges. Furthermore, parents who apply for physical aid on behalf of their children often feel a loss of autonomy. Autonomy is a crucial principle when making decisions about children's health [32]. However, these parents are limited to phone communication and lack digital access to their child's information, which can be both frustrating and challenging for them.

Findings from this study show that a third-party play a crucial role in reducing administrative burdens for vulnerable citizens who struggle with digital self-services by "imposing" their privacy and conducting workarounds. This highlights the importance of designing user-friendly digital self-services that consider the struggles of vulnerable citizens who rely on third-party assistance and face administrative burdens. Despite being designed to protect sensitive information, personal digital devices have become a source of burden for vulnerable citizens, creating a need to address privacy dimensions to better serve these citizens.

4.3 Learning Costs – Documentation Issues

Learning costs describe the volume of information citizens must comprehend regarding the service to determine their eligibility and understand the process [46]. Previous research says that public services are primarily designed from an administrative standpoint, and there is typically a gap between the government and the citizen's perception of necessary information, understanding, and documentation requirements [28] This study aligns with this, as findings show that vulnerable citizens have significant challenges when it comes to submitting documents and removing unnecessary sensitive information when uploading digitally which they are required to in order for their application to be accepted. In addition, vulnerable citizens' privacy regulations make sharing and reuse of documents within the organization difficult.

Vulnerable citizens applying for local services experience higher documentation demands then citizens applying for central services because the organizations software and employees cannot share documents and information regarding local services between them. The process of submitting documentation for financial support presents a significant challenge for vulnerable citizens, often resulting in frustration with self-service applications. This challenge is particularly overwhelming for first-time applicants, as NAV requires all their personal finances to be documented to make the right discretion for citizens who apply.

When citizens submit their financial overview for the last three months, many include unnecessary personal information in their bank transactions overview. Although citizens can delete sensitive information, many are unaware of this option or simply overlook it. Additionally, some citizens prefer to submit paper documentation at the local NAV office, despite the online application option, as they feel more reassured when a frontline workers can verify the correctness of their documents.

However, this preference for paper submissions leads to inadvertent exposure of sensitive information, such as bank transactions, on the shared printers or computers in the self-serving area at the local NAV office. Sensitive information is often stored on computers at local NAV offices without proper deletion protocols. Similarly, citizens seeking assistance to upload documents online at the local NAV office may unintentionally expose their personal information to third parties exposing them for identity theft and fraud. Moreover, if frontline workers are unavailable, citizens often seek assistance from anyone nearby, further compromising their privacy. Overall, many citizens struggle to collect and submit the necessary documents independently, often requiring assistance to navigate this complex process.

Furthermore, while previously submitted documentation can be reused for new and other applications, citizens face challenges when attempting to reuse documentation from other central services due to NAV's lack of cross-service documentation sharing capabilities. Usually, parents with chronically ill children apply for various services through NAV, including financial support for children or adults requiring long-term private care, income during absences from work due to child hospitalization or treatment, and additional financial assistance for permanent injuries, illnesses, or disabilities. However, when applying for Physical Aid they must provide documentation of their child's illness once again, despite NAV already having this information from other applications. This duplication arises because caseworkers at the NAV Physical Aid center lack access to documentation stored in central government software. As a result, parents often resort to repeatedly contacting their children's general practitioners to obtain documentation, which is perceived as an administrative burden. This is especially true for children facing ongoing health challenges with no signs of improvement, as illustrated by one participant who remarked *"no, the arm still hasn't grown back"* (OB20 – Parent of chronical ill/disabled children) This process creates administrative burdens and may be perceived as confusing for citizens who do not understand that the same organization is divided into sections with limited access to each other.

The cognitive load placed on citizens by requiring them to manage more information and make complex choices can lead to feelings of overwhelm and result in higher administrative burdens [51]. Despite these challenges, citizens are often tasked with protecting their personal information, even when they feel powerless to do so effectively [52]. Vulnerable citizens encounter additional learning costs related to privacy regulations with digital self-service platforms. When applying for financial support, they must navigate the process of determining which documents to provide and how much personal information to disclose, adding to their administrative burdens. Moreover, duplication of documentation requirements and reliance on third parties further compound these challenges.

Privacy is widely acknowledged as crucial for personal autonomy and dignity [15, 18] However, there are instances where individual privacy rights may be overridden by the public interest in disclosing personal information, necessitating a delicate balance between competing interests [15]. Balancing privacy regulations with the information necessary for caseworkers to make informed benefit decisions requires careful consideration of what constitutes "reasonable" documentation. The determination of this balance seems to be context-dependent and varies across services underscoring the complexity of the issue addressed in this research. These findings suggest that vulnerable citizens prioritize anticipated benefits over privacy regulations when applying for benefits, highlighting the need for nuanced approaches to address privacy regulations in self-services systems.

5 Conclusion and Future Research

Using qualitative methods to explore citizens' experiences of administrative burdens, this study has shown that vulnerable citizens do not perceive digital self-services as user-friendly due to the challenges posed by privacy regulations. Furthermore, digitalizing welfare contributes to growing inequality in the distribution of burdens among citizens, reproducing socioeconomic inequalities [26, 28, 53]. This study was guided by the research question: *What challenges of privacy regulations related to digital self-services do vulnerable citizens experience?* The findings provide several examples on how the government is adding administrative burdens on the most vulnerable citizens through policy design and bureaucratic dysfunction [19, 34] when providing welfare services on digital self-service platforms.

This study contributes to the theoretical understanding of the link between digitalization and administrative burden by utilizing the typology of costs [18] to identify the administrative burdens experienced by vulnerable citizens in Norway when applying for financial support or physical aids through digital self-services. The strategy of digitalizing welfare services has ironically increased administrative burdens for vulnerable citizens in various ways, despite NAV's goal of protecting citizens' privacy [13]. Their one-stop-shop approach and digital-first strategy for digital self-services have contributed to these burdens. [25, 39]. The study also supports prior research demonstrating how citizens can be restricted to using specific channels without alternative options and become dependent on gatekeepers to utilize benefits [4, 35]. However, this study looks deeper into the role of privacy regulations and present it as one of the reasons for why citizens experience these communication issues.

Privacy restrictions result in compliance costs, limiting access to information and assistance, while the agency channel strategy's lack of support for vulnerable citizens creates psychological costs such as the risk of fraud and identity theft. Further, there are learning costs as documentation demands can be difficult for vulnerable citizens to meet, and limited opportunities for reusing documentation create additional burdens for those who need to resubmit documents.

This study addresses the need for research that sheds light on the issues that contribute to the unfair distribution of administrative burdens [16, 22]. These burdens can lead to vulnerable citizens experiencing scarcity, increased stress, and feelings of overwhelm, potentially contributing to burnout with government programs [19, 31, 32, 54]. Furthermore, these challenges reinforce social inequalities for vulnerable citizens who rely on welfare services, highlighting the need for the government to address these differences. Additionally, this study highlights the critical need vulnerable citizens have for assistance from third-party actors as they are left in the dark and rely on others to utilize self-services [47, 48]. However, this situation raises concerns about how vulnerable citizens are often "forced" to disclose their private information, financial status, and health issues throughout the process when receiving assistance and the study raises questions about what citizens without a support system or network can do in these situations. Future research should explore the role of these third-party actors while considering the administrative burdens they may experience when assisting someone while standing in the dilemma of revealing privacy or deciding not to help. Additionally, investigating the strategies and workarounds employed by both citizens and welfare workers to ease administrative burdens can provide valuable insights for re-designing and creating more user-friendly digital self-services. These workarounds also indicate that these burdens are accidental, and that welfare workers feel that they cannot change government actions [28]. The firsthand experiences of citizens shared in this study can provide policy makers with a better understanding of how government designs for the mainstream are perceived as burdensome by vulnerable citizens.

Acknowledgments. This research for this paper was co-financed by the Norwegian Labour and Welfare Administration (NAV) and the Research Council of Norway [project number 316246].

References

1. Falk, S., Römmele, A., Silverman, M.: Digital Government. Springer, Cham (2017)
2. Breit, E., et al.: Digital coping: how frontline workers cope with digital service encounters. Soc. Policy Admin. **55**(5), 833–847 (2021)
3. Pors, A., Schou, J.: Street-level morality at the digital frontlines: an ethnographic study of moral mediation in welfare work. Admin. Theory Praxis **43**(2), 154–171 (2021)
4. Lindgren, I., et al.: Exploring citizens' channel behavior in benefit application: empirical examples from Norwegian welfare services. In: Proceedings of the 15th International Conference on Theory and Practice of Electronic Governance (2022)
5. Lindgren, I., et al.: Close encounters of the digital kind: a research agenda for the digitalization of public services. Gov. Inf. Q. **36**, 427–436 (2019). https://doi.org/10.1016/j.giq.2019.03.002
6. Bataineh, E., B. Al Mourad, and F. Kammoun. Usability analysis on Dubai e-government portal using eye tracking methodology. in 2017 Computing conference. 2017. IEEE
7. Galvez, R.A., Youngblood, N.E.: E-Government in Rhode Island: what effects do templates have on usability, accessibility, and mobile readiness? Univ. Access Inf. Soc. **15**(2), 281–296 (2016)
8. Huang, Z., Benyoucef, M.: Usability and credibility of e-government websites. Gov. Inf. Q. **31**(4), 584–595 (2014)
9. Islam, M.N., Rahman, S.A., Islam, M.S.: Assessing the usability of e-government websites of Bangladesh. In: 2017 International Conference on Electrical, Computer and Communication Engineering (ECCE). IEEE (2017)
10. Venkatesh, V., Hoehle, H., Aljafari, R.: A usability evaluation of the Obamacare website. Gov. Inf. Q. **31**(4), 669–680 (2014)
11. Lyzara, R., et al.: E-government usability evaluation: Insights from a systematic literature review. In: Proceedings of the 2nd International Conference on Software Engineering and Information Management (2019)
12. Shneiderman, B.: Universal usability: pushing human–computer interaction research to empower every citizen, in Media Access, pp. 275–286, Routledge. (2003)
13. Joshi, J.B.D., et al.: Digital government security infrastructure design challenges. Computer **34**(2), 66–72 (2001)

14. Jansson, G., Erlingsson, G.Ó.: More e-government, less street-level bureaucracy? On legitimacy and the human side of public administration. J. Inform. Tech. Polit. **11**(3), 291–308 (2014)
15. Westin, A.F.: Privacy and freedom. Washington Lee Law Rev. **25**(1), 166 (1968)
16. Madsen, C.Ø., Lindgren, I., Melin, U.: The accidental caseworker–how digital self-service influences citizens' administrative burden. Gov. Inf. Q. **39**(1), 101653 (2022)
17. Kotamraju, N.P., van der Geest, T.M.: The tension between user-centred design and e-government services. Behav. Inf. Technol. **31**(3), 261–273 (2012)
18. Herd, P., Moynihan, D.P.: Administrative burden: Policymaking by other means. Russell Sage Foundation (2019)
19. Zhao, J., Tomm, B.M.: Psychological responses to scarcity, in Oxford research encyclopedia of psychology (2018)
20. Heinrich, C.J., Brill, R.: Stopped in the name of the law: administrative burden and its implications for cash transfer program effectiveness. World Dev. **72**, 277–295 (2015)
21. Barnes, C.Y., Henly, J.R.: "They are underpaid and understaffed": how clients interpret encounters with street-level bureaucrats. J. Public Admin. Res. Theory **28**(2), 165–181 (2018)
22. Chudnovsky, M., Peeters, R.: The unequal distribution of administrative burden: a framework and an illustrative case study for understanding variation in people's experience of burdens. Soc. Policy Admin. **55**(4), 527–542 (2021)
23. Hansen, H.T., Lundberg, K., Syltevik, L.J.: Digitalization, street-level bureaucracy and welfare users' experiences. Soc. Policy Admin. **52**(1), 67–90 (2018)
24. Kim, E., Lee, B., Menon, N.M.: Social welfare implications of the digital divide. Gov. Inf. Q. **26**(2), 377–386 (2009)
25. Baskerville, R.L., Myers, M.D., Yoo, Y.: Digital first: the ontological reversal and new challenges for IS research. MIS Q. **2020**(44), 509–523 (2019)
26. Ranchordás, S.: The digitization of government and digital exclusion: setting the scene. In: Blanco de Morais, C., Ferreira Mendes, G., Vesting, T. (eds.) The Rule of Law in Cyberspace. Law, Governance and Technology Series, vol. 49, pp 125–148. Springer, Cham (2022). https://doi.org/10.1007/978-3-031-07377-9_7
27. Dodds, S.: Dependence, care, and vulnerability. Vulnerability: new essays in ethics and feminist philosophy, pp. 181–203 (2014)
28. Herd, P., et al.: Introduction: administrative burden as a mechanism of inequality in policy implementation. RSF: Russell Sage Found. J. Soc. Sci. **9**(4), 1–30 (2023)
29. Nisar, M.A.: Children of a lesser god: administrative burden and social equity in citizen–state interactions. J. Public Admin. Res. Theory **28**(1), 104–119 (2018)
30. Heggertveit, I., Lindgren, I., Madsen, C.Ø., Hofmann, S.: Administrative burden in digital self-service: an empirical study about citizens in need of financial assistance. In: Krimmer, R., et al. (eds.) EPart 2022. LNCS, vol. 13392, pp. 173–187. Springer, Cham (2022). https://doi.org/10.1007/978-3-031-23213-8_11
31. Lindström, C., Åman, J., Norberg, A.L.: Increased prevalence of burnout symptoms in parents of chronically ill children. Acta Paediatr. **99**(3), 427–432 (2010)
32. Batchelor, L.L., Duke, G.: Chronic sorrow in parents with chronically ill children. Pediatr. Nurs. **45**(4), 163–183 (2019)
33. Vonneilich, N., Lüdecke, D., Kofahl, C.: The impact of care on family and health-related quality of life of parents with chronically ill and disabled children. Disabil. Rehabil. **38**(8), 761–767 (2016)
34. Christensen, J., et al.: Human capital and administrative burden: the role of cognitive resources in citizen-state interactions. Public Adm. Rev. **80**(1), 127–136 (2020)
35. Collyer, F.M., Willis, K.F., Lewis, S.: Gatekeepers in the healthcare sector: knowledge and Bourdieu's concept of field. Soc Sci Med **186**, 96–103 (2017)

36. Larsson, K.K.: Digitization or equality: When government automation covers some, but not all citizens. Gov. Inf. Q. **38**(1), 101547 (2021)
37. Nielsen, M.M.: The untapped potential: the inclusive, personal and co-created public service experience in Europe. In: Musiał-Karg, M., Luengo, Ó.G. (eds.) Digitalization of Democratic Processes in Europe. Studies in Digital Politics and Governance, pp. 165–188. Springer, Cham (2021). https://doi.org/10.1007/978-3-030-71815-2_13
38. Madsen, J.K., Mikkelsen, K.S., Moynihan, D.P.: Burdens, sludge, ordeals, red tape, oh my!: a user's guide to the study of frictions. Public Admin. **100**(2), 375–393 (2022)
39. Askim, J., et al.: One-stop shops for social welfare: the adaptation of an organizational form in three countries. Public Admin. **89**(4), 1451–1468 (2011)
40. Finne, J., Sadeghi, T., Løberg, I.B., Bakkeli, V., Sehic, B., Thørrisen, M.M.: Predictors of satisfaction with digital follow-up in Norwegian Labor and welfare administration: a sequential mixed-methods study. Soc. Policy Admin. **57**(7), 1150–1165 (2023)
41. Fugletveit, R., Lofthus, A.-M.: From the desk to the cyborg's faceless interaction in the Norwegian labour and welfare administration. Nordic Welf. Res. **6**(2), 77–92 (2021)
42. Parker, A., Tritter, J.: Focus group method and methodology: current practice and recent debate. Int. J. Res. Method Educ. **29**(1), 23–37 (2006)
43. Nyumba, O.T., Wilson, K., Derrick, C.J., Mukherjee, N.: The use of focus group discussion methodology: Insights from two decades of application in conservation. Methods Ecol. Evol. **9**(1), 20–32 (2018)
44. Braun, V., Clarke, V.: Using thematic analysis in psychology. Qual. Res. Psychol. **3**(2), 77–101 (2006)
45. Krippendorff, K.: Content Analysis: An Introduction to its Methodology. Sage publications, Thousand Oaks (2018)
46. Moynihan, D., et al.: Matching to categories: learning and compliance costs in administrative processes. J. Public Admin. Res. Theory **32**(4), 750–764 (2022)
47. Hoglund Ryden, H., De Andrade, L.: The hidden costs of digital self-service: administrative burden, vulnerability and the role of interpersonal aid in Norwegian and Brazilian welfare services. In: Proceedings of the 16th International Conference on Theory and Practice of Electronic Governance (2023)
48. Alshallaqi, M., Al-Mamary, Y.H.: Paradoxical digital inclusion: the mixed blessing of street-level intermediaries in reducing administrative burden. Gov. Inf. Q. **41**(1), 101913 (2024)
49. Madsen, C.Ø., Kræmmergaard, P.: Channel choice: a literature review. In: Tambouris, E., et al. EGOV 2015, vol. 9248, pp. 3–18. Springer, Cham (2015). https://doi.org/10.1007/978-3-319-22479-4_1
50. Buffat, A.: Street-level bureaucracy and e-government. Public Manag. Rev. **17**(1), 149–161 (2015)
51. Schwartz, B.: The tyranny of choice. Sci. Am. **290**(4), 70–75 (2004)
52. Bekker, S.: Fundamental rights in digital welfare states: The case of SyRI in the Netherlands. Netherlands Yearbook of International Law 2019: Yearbooks in International Law: History, Function and Future, pp. 289–307 (2021)
53. Halling, A., Baekgaard, M.: Administrative burden in citizen-state interactions: a systematic literature review. J. Public Admin. Res. Theory **34**(2), 180–195 (2024)
54. Hilbert, L.P., Noordewier, M.K., van Dijk, W.W.: The prospective associations between financial scarcity and financial avoidance. J. Econ. Psychol. **88**, 102459 (2022)

Mitigating Administrative Burdens: Understanding the Role of Intermediaries in Co-producing Digital Self-services

Hanne Höglund Rydén[1]([✉])[iD], Sara Hofmann[1][iD],
Luiz Henrique Alonso de Andrade[2][iD], and Ida Heggertveit[1][iD]

[1] University of Agder, Kristiansand, Norway
hanne.s.h.ryden@uia.no
[2] University of Tampere, Tampere, Finland

Abstract. Public sector organizations are increasingly adopting digital self-service solutions. With digital self-services, citizens are given a more active role in co-producing their services, as they serve themselves and perform tasks that were previously performed by public professionals. However, many citizens struggle with their expanding role as co-producers and experience burdens in their interaction with digital self-services. To mitigate these burdens, citizens often turn to so-called intermediaries for help. Intermediaries are third parties such as family members, friends as well as professionals that assist citizens during digital self-services or completely take over the responsibility from them. Although they are important co-producers, intermediaries have seldom been the focus of attention as their role during co-production is often invisible from the outside. We present two qualitative empirical studies from Norwegian and Brazilian welfare services. Our findings show the burdens citizens experience with digital self-services and illuminate how important personal intermediaries are to reduce citizens' experience of burdens, resulting in triangulated or hybrid "co-production".

Keywords: Digital self-services · co-production · administrative burdens · intermediaries

1 Introduction

Global policies frequently promote the effectiveness of Information and Communication Technologies (ICT) within the framework of e-government or digital government. These technologies revolutionize the way citizens interact with the public sector, with digital self-services replacing in-person service delivery as citizens interact directly with a system [2]. Self-services have the potential to lower costs in public administration and enable service co-production, where citizens complete some of the administrative work to receive the service they require [3, 4].

A central element of digital self-services is the idea of co-production, where both the service consumer (citizens) and service provider (public sector) are involved in the delivery of the service [5]. Co-production has always been an important prerequisite for the delivery of public services as citizens have always been required to deliver some sort of input such as handing in documents, thus contributing to, and shaping the outcome of a public service. However, with digital self-services, citizens are given a more active role in co-producing their services [4] as they serve themselves and perform tasks that were previously performed by public professionals [1].

When citizens experience burdens with digital self-services, they often seek the assistance of intermediaries to overcome these burdens and to co-produce the self-services. Intermediaries can be understood as third-party actors that facilitate and coordinate interactions between parties in service encounters [10, 11]. Intermediaries can be individuals or organizations such as public libraries, acting as intermediaries between government services and citizens in need of support [6]. Another example of intermediaries includes family members and friends with experience in dealing with the public sector who provide support based on their own knowledge of services [12].

Previous research on co-production in digital self-services has primarily focused on the interactions between citizens, the public sector, and technology [8]. However, the role of intermediaries has started to gain recognition as essential actors and co-producers in digital self-services, as they can help reduce citizens' burdens [13]. This paper answers the call for more qualitative research that explores the nature of burdens from the perspective of citizens, and how actors outside the classical understanding of co-production influence citizens experience of administrative burden [21]. We explore this by comparing burdens across welfare contexts, addressing the following question: ***What is the role of intermediaries for mitigating the burdens citizens experience when co-producing digital self-services?***

To answer this question, we draw on relevant literature that addresses digital self-services, co-production, and intermediaries in the context of digital government. We present empirical data collected from two qualitative case studies conducted in similar social security agencies, one in the Norwegian Labor and Welfare Administration (NAV) and another in the Brazilian National Social Security Institute (INSS). Our analysis is framed by the concept of administrative burdens, which defines the various costs associated with administrative procedures and explores how individuals with different preconditions, cope with these burdens across the two diverse welfare contexts [9, 11, 14, 15].

The paper is structured as follows: We start by discussing the literature that provide our theoretical background linking the concepts of co-production, administrative burdens, intermediation to self-services in digital government. Next, we present our research design and empirical context followed by findings and analysis. Finaly, we discuss the role intermediaries can play for citizens experiences of administrative burdens in co-producing digital self-services as well as limitations and suggestions for further research.

2 Theoretical Background

In this section, we present the concepts of co-production in digital self-services, the administrative burdens citizens perceive while co-producing digital self-services, as well as how intermediaries help citizens during co-production. The literature highlights presented in this section provide the theoretical linkages between these concepts and thus the background for the analysis of our empirical data.

2.1 Co-production in Digital Self-services

Digital self-services shift the responsibility of task completion from public sector employees to citizens, transforming them into 'screen level workers' in charge of managing their own service procedures [2]. Citizens are even referred to as the new employees, highlighting the changed nature of their role [4]. As a result, digital self-services also alter the roles of citizens and frontline workers, creating novel dynamics for co-production [8, 15].

Co-production refers to the involvement of customers in the production of services, with co-production being an integral part of the service process that cannot be separated from the service itself [28]. In line with this understanding, we view digital self-services as inherently linked to co-production [5]. There has been an ongoing discussion in various literature streams about which actors should be considered part of the co-production of self-services and the level of citizen engagement in this process [1]. In the field of public administration and management, co-production has been described as an exchange between the public sector and clients, aiming to enhance government effectiveness by actively involving citizens in the service process [3]. The motivation behind co-production is often driven by economic reasonings, allowing for the allocation of work when government resources are limited [3, 29].

Self-service policies promise advantages for both citizens, who become empowered co-producers as they can take an active role and access services online from home, as well as public sector organizations, which save resources [30, 31]. In addition, the bypassing of physical encounters between citizens and government has been argued to remove burdens associated with office visits and meetings as services are co-produced digitally [32]. Besides, by fostering a more responsible and engaged citizenry that takes an active role in service co-production, financial burdens on citizens' community can be reduced as relocating work to citizens will get better value out of government [33].

2.2 Administrative Burdens in Co-production of Digital Self-services

While self-service co-production can eliminate burdens and provide opportunities for empowerment, it also carries a risk of placing an excessive burden on citizens [9, 32]. When citizens enroll in the public service procedure, they often face a rigid, complex and hard-to-navigate public sector that expects service applicants to comply with a multitude of forms, regulations, and decisions. These elements have been framed in the literature as administrative burdens when the public encounter imposes transaction costs on citizens that need to spend their resources to access public services [14, 34].

Administrative burdens refer to the costs, in terms of time, effort, and resources, that individuals or organizations face when they interact with public administration or government processes. These burdens can be broken down into three main categories [14, 34]. *Compliance costs* refer to the resources that citizens spend to ensure adherence to regulations and procedures set by government agencies. They include individual expenses related to filling out forms, obtaining necessary documentation, or meeting specific criteria mandated by government such as traveling to attend meetings as well as time to collect documentation. *Learning costs* refer to the time and effort invested in research and understanding the relevant laws, procedures, and bureaucratic processes to ensure compliance. *Psychological costs* refer to the emotional and mental toll that administrative burdens can have on individuals, such as feelings of powerlessness or frustration when facing bureaucratic and hard to navigate services. They are often associated with waiting times, unclear communication, and perceived bureaucratic complexities that impact the overall experience. When citizens attempt to navigate complex bureaucratic processes that involve these costs, it can lead to energy loss, expenses, frustration, and anxiety. These burdens will negatively impact individuals' mental health and well-being and can lead to diminished trust in governmental institutions and authorities [14, 34]. Furthermore, they can impact citizens' access to fundamental rights and often play a role in reinforcing existing inequalities [14].

Table 1. The costs of administrative burdens (Herd and Moynihan, 2019; p.23)

Type of cost	Description of costs
Learning costs	Learn about and understand the program or service, ascertaining eligibility, the benefits nature, conditions that must be met, and how to gain access
Compliance costs	Provide information and documentation to demonstrate standing; financial costs to access services (fees, legal representation, travel costs); avoiding or responding to administrators' discretionary demands
Psychological costs	Frustration from intrusive administrative supervision and dealing with learning and compliance costs, unjust or unnecessary procedures; stress from uncertainty about whether negotiating the processes is possible

Public organizations often implement digital technologies that aim to reduce burdens of bureaucracy within the organization [35], but less focus has been spent on understanding citizens experiences of burdens [6, 9, 21]. In order to mitigate the burdens, they face, citizens make use of support mechanisms, such as the reliance on a third person's personal help [14]. The social infrastructure can help mediate burdens born from digitalization, making personal support vital for citizens to cope with administrative burdens [6]. However, as of now, the role of these intermediaries in mitigating the burdens citizens experience in the co-production of digital self-services has not been comprehensively investigated.

2.3 Intermediaries in the Co-production of Digital Self-services

Intermediaries are important actors in the service procedure as these actor's bridge interactions by matching needs (demand) with services (supply), gather information, provide support and trust throughout the process to assure decision-making accuracy and facilitate interactions [36]. Intermediaries can broadly be defined as "any public or private organization facilitating the coordination between public service providers and their users" [6] or as brokers that facilitate knowledge exchange between actors [11]. In such a sense intermediaries can ease complexity in government transactions and reduce costs and the administrative burden citizen experiences [10].

There is currently no universally accepted definition that fully encompasses the concept of 'intermediary'. However, in the digital government literature, intermediaries are often described as access points to connect individuals with a complex system. For instance, private or public organizations may serve as these access points, acting as bridges between citizens and digital government. Similarly, civil society intermediaries can facilitate interactions that help citizens engage with government and its services [38]. Civil society intermediaries can be family members and friends that accompany citizens in their interactions with government, such as parents that assist their children in applying for unemployment benefits after they have finished school. In relation to public sector digitalization, intermediation can be performed by acquaintances—"warm experts"— who are knowledgeable in administrative procedures and/or the digital domain, dealing with the digital service in question in favor of the another citizen [12].

Digitalization enables new kinds of intermediation practices by providing platforms that mediate online interaction where actors come together to co-produce services [38]. For example, the Norwegian public sector platform myNAV provides a logged in site where unemployed citizens and work counselor can interact to communicate goals, discuss progress, answer each other's questions, and share information, which brings the participants closer to a common goal as services are co-produced. However, in the contexts of welfare services, technology has been argued to change public service provision in a way that some citizens experience burdensome [15]. This makes the assistance from human actors essential for citizens to co-produce services [6].

These perspectives on intermediaries position them across individual (e.g. social network), organizational (e.g. professional experts) and technical dimensions (e.g. public/private platforms) as actors in service co-production. By addressing technology as an intermediary, the digital becomes visible as an actor in the co-production of digital self-services, which has been argued important in previous studies [8].

Organizational intermediaries can also be positioned outside the public service organization, in partnerships guided by formal rules as they support and empower users in the process of service co-production [7]. In such an understanding, intermediaries are actors separated from the public service organization, interacting with the government in favor of another party [7]. This can, for example, be a private organization intermediary specialized in tax law and procedures that administers the declarations in favor of a private business. Depending on the context, intermediaries can ease the administrative burdens citizens face in digital service encounters, but they can also impose new burdens on citizens when power is relocated from citizens to a third-party intermediary [13]. This can be the case if intermediating agencies or brokers benefit from services that make citizens experience scarcity and financial constraints. The intermediary can also misuse power leverage in relation to the person they help, for example, in a domestic violence relation where the abuser takes increased control over the abused by intermediating their affairs with public agencies and private businesses.

Although intermediaries take different forms in different settings and their roles can vary in digital government service delivery, their common practice is to provide help that benefits both citizens and service organizations [37]. This make actors outside the public sector organization important contributors in digital self-service procedures [6] as co-producers of such procedures [8]. In this paper, we frame intermediaries out of two dimensions, personal intermediaries—humans—and digital intermediaries—technology [38]. Digital or electronic intermediaries' position in a digital government usually means publicly owned websites, portals or apps that provide information about government activities or access to services, offering citizens information and help that they might find valuable [39]. Both the digital and human intermediaries facilitate interaction between parties, and we understand them from the broader sense as being actors that can be positioned both inside and outside public sector. In this sense, they provide access points to connect individuals with the digital government system [38].

3 Research Design and Empirical Context

In this section, we present our empirical context that is rooted in two qualitative cases studies of self-service practices in the Norwegian and Brazilian welfare states [25]. The two case studies were inspired by multi-sited ethnography [16]. NAV and INSS have been avantgarde, leading organizations in terms of digitalization in their countries. Around 2019, both provided fully-fledged digital self-services, mainly through backend database interoperability and frontend online portals (myNAV) and cellphone apps (myINSS) [17–20]. While Norway, as a Nordic country, is considered an example of a social-democratic, universal welfare state [19, 20], Brazilian welfare arrangements, like those in other South American democracies, provide in most cases 'hybrid', two-tier protection systems: underfunded universalistic policies, and privately paid—yet state-sponsored—improved welfare services [17, 18]. Although state reform waves progressively align social policies in both contexts according to general western neoliberal trends, e.g., introduction of selective instruments like means-testing and activation policies [17], the welfare systems and contexts are still fundamentally different, if not diametrically opposed—a fact we draw from to increase external validity of the shared case findings [25, 27]. Yet, despite a former huge gap between the countries, Norway and Brazil share similar scores in terms of digital service provision, according to the United Nations' E-Government Knowledge Base [22], and both agencies manage comprehensive portfolios of welfare policies [23, 24], which include cash-based, means-tested social assistance benefits—whose entitlement depends on citizens meeting income or vulnerability criteria. The welfare policies that define these criteria involve complex procedures—and thus administrative burden become barriers for citizens in their self-service procedure. The two cases were selected to understand how digitalization of welfare services, in particular the implementation of self-service technology [4], impacts citizens' ability to co-produce their services in different welfare contexts. While operating similar welfare systems, through similar levels of digital services, NAV and INSS are located far from each other in the welfare spectrum—especially context-wise and policy-wise—thus conferring reliability to our findings [25, 27].

The data collection focused on citizens' encounters with the digitalized welfare state and the role of intermediaries in the digital self-service co-production. Fieldwork was pursued in both cases by observations and interviews with citizens that interacted with self-service technology in different social security service offices of the two countries' national social security agencies (NAV and INSS) [25]. The citizens we encountered had experience with using digital public systems on their personal devices or by using self-service stations provided by the offices. Data collection took place in 2022 and 2023. We interviewed citizens who had applied or wanted to apply for social assistance benefit, 12 (lasting between 5 and 45 min each) in two NAV offices, and 12 (lasting between 5 and 20 min each) across three INSS offices. In total, 8 h and 35 min were invested in NAV office observations and 11 h and 20 min in INSS. Observations were conducted at both offices as citizens interacted with digital self-service stations. We also took fieldnotes from conversations with public officials at the offices whose additional perspectives gave further insight into how citizens managed to co-produce using self-services INSS and NAV provided. Altogether, 24 interviews, 24 observations and five days of fieldnotes from informal office conversations were transcribed, coded, and analyzed in NVivo through theory informed content analysis [26]. Since our data collection focused on those citizens that were physically present at the offices, this increased the odds that they did face some kind of burden—out of the assumption that they otherwise would be able to manage their cases without attending a service office. We understand these citizens as digitally disadvantaged, as they seldom were favored by the digital services and seldom desired to engage in digital encounters [13].

The case choice was convenient, as the authors were familiar with the two welfare contexts under study: this made it possible to conduct a multiple case study valuable to exploring new topics in similar settings [27]. Multiple case studies are especially valuable because they provide greater analytical strength for the theoretical relationships proposed when similar patterns are observed across different cases [27].

To derive valid inferences from the two data sets and identify the cost rising from administrative burdens in self-service and the role of intermediaries in bridging or enforcing them, we conducted a deductive qualitative content analysis. During this analysis, we also allowed for inductive themes to emerge from the data analysis, such as the different categories of intermediaries. The process involved open coding, where we took notes in the text while reading it. We then read the text again and documented descriptions of the content. Subsequently, we grouped these descriptions into new categories in order to find similarities between the two cases and provide a more comprehensive description of the phenomenon of co-production in digital self-service [26].

4 Intermediaries' Role in Mitigating Administrative Burdens for Citizens

In this section, we provide empirical examples of the three different costs (see Table 1) of administrative burden faced by citizens in our two cases of digital self-service co-production. In our cases, the digital intermediaries are myNAV in Norway and myINSS in Brazil. Personal intermediaries include frontline welfare office workers in both cases, but also paid brokers in Brazil called Despachantes. In Table 2 we summarize the administrative burdens and costs in relation to personal and digital intermediaries. The analysis is based on the interpretation of interviews and observations with citizens, under the light of field notes on informal conversations with frontline workers. We include interviewees' quotes to illustrate and support the findings.

Table 2. Summary of findings originated from observations and interviews

Costs		Norway (NAV)	Brazil (INSS)
Learning Cost (LC)	Personal intermediaries	Frontline workers can *reduce LC* by acting on behalf of citizens. However, this may not enable citizens to learn, and they may require assistance again in the future	Using external intermediaries can *reduce the LC* by acting on behalf of citizens. However, this approach may not enable citizens to learn, and they may require assistance in the future as well
	Digital intermediaries	Information is provided on the webpage; however, it is complex to understand, in fact often *worsening LC*	Complexity and mismatching information (e.g., different service names) can confuse citizens and compromise learning. Laws and regulations are hard to understand through digital means, *worsening LC*
Compliance Costs (CC)	Personal intermediaries	Frontline workers are not supposed to help citizens using self-services, but they do if they have the time and then *reduce CC* by performing task for citizens	Despachantes reduce CC by assisting citizens in utilizing self-services for a fee. They also impose a 'financial' CC by charging citizens for their services
	Digital intermediaries	The webpage is often too complex for citizens to understand, *worsening CC*	The technology does not always work or work as expected, and the lack of functionality *worsen CC*

(*continued*)

Table 2. (*continued*)

Costs		Norway (NAV)	Brazil (INSS)
Psychological Costs (PC)	Personal intermediaries	Frontline workers with restricted resources create stress for citizens, *worsening PC*. When they confirm that citizens have rightly conducted self-service, *PC are reduced*	Frontline workers sometimes treat citizens disrespectful and impose stigmata, *worsening PC*. Paid intermediators (Despachantes) acquainted with laws and procedures *reduce PC* by easing their stress
	Digital intermediaries	The "digital first" strategy "forces" citizens to apply digitally, creating a feeling of loss of autonomy, *worsening PC*	Unstable technology, rule and information complexity create uncertainty and stress that *worsen PC*

4.1 Learning Costs

In this section, we discuss learning costs that relate to time and effort citizens in NAV and INSS experience in their digital self-services and how these learning costs are reduced or enforced by citizens encounters with a) digital intermediaries and b) personal intermediaries in self-service co-production.

The Role of Digital Intermediaries in Creating Learning Cost

Extensive information is available in both INSS's and NAV's digital systems. However, citizens in NAV struggle to navigate the huge amount of information provided and to understand how this information is relevant to their case as the digital intermediary of information does so in a complex bureaucratic language: *"There is a lot of information available at NAV.no, but it is hard to understand it and to know what information that is right for me, that is the right one in my situation. It takes time to understand and then you give up"*. In INSS, struggling citizens—elderly in special—often face challenges in understanding how to use myINSS. Instead of easing, the digital self-service increases their learning costs. Further, many citizens repeatedly seek assistance at the office as they need to relearn the digital procedures whenever they must deal with them, creating enduring learning costs. One person told us: *"The physical unit should always exist because the situations are complex, and the bureaucratic language, people do not understand regardless of education"*. In another case, a citizen asked for guidance about a service as she could not find it via myINSS, because the very name of the service in the app was unusual, preventing understanding. Some citizens describe that the learning costs of using the digital platform and learning about the services in that way becomes more challenging and imposes additional costs: *"I looked for information on the Internet, but it is not easy to find, the rules are too confusing. I spent half an hour reading and there was no use"*. Therefore, the digital intermediaries can impose extra and *enduring* learning costs, especially for citizens that already struggle with digital self-services.

The Role of Personal Intermediaries in Mitigating or Enforcing Learning Cost

Frontline workers in NAV often took over the self-service procedure and navigated the platform on citizens' behalf. They asked questions and filled out the information requested by the system while the citizen became a bystander. This bridged the learning costs for the citizens as the intermediary took over responsibility and often knew how to go about fast and efficiently. However, this only bridged the learning cost at the specific occasion as one citizen told us: *"Next time I will come to the office again, to get assistance, like today, I will not remember how to go about on my own. I have not learned it. I always come to the office to get help every time I need to apply"*. While the intermediary mitigated the learning costs for the time being, the citizen failed to understand the procedure to be able to later engage in the digital self-service independently. Hence, the same enduring learning cost is imposed by NAV as by INSS, when citizens attend offices to "learn" how to use the digital services on and again.

Our observations and interviews also showed that the public officials in INSS offices seldom help citizens in their requests, instead directed them to myNSS not taking imposed learning costs from them. One citizen described that *"They (INSS officials) didn't even tell me what documents I must present. Probably my Work Register Booklet ("Carteira de Trabalho") to prove I have no income."* To bridge these costs, in Brazil, intermediaries outside of the public sector help citizens to formulate the right questions to receive adequate responses from INSS officials and guide citizens to use myINSS. Some citizens receive support from family members and people in their network to help bridge learning costs. Citizens that do not have intermediaries to help them in understanding the regulations, are left in the dark: *"There is a lot of bureaucracy, a lot of paperwork, nobody clarified it properly. The doctor asked why they (INSS) rejected (my application). I could not answer."*

4.2 Compliance Costs

This section discusses compliance costs of potential time and money citizens spend in their government encounters and how this cost is enforced with a) digital intermediaries and reduced with b) personal intermediaries in self-service co-production.

The Role of Digital Intermediaries in Creating Compliance Costs

In Norway, citizens where often asked to use the self-service stations at the office. Many citizens struggled to use these stations, understanding the written information, navigating information and applications, and filling out their information in the digital application. The fear of making mistakes prevented some from applying for services in cases where no help was provided, and some citizens left the office without applying at all. In these cases, the myNAV digital intermediary failed to reduce burdens as it was experienced as too complex for the citizens. Instead of having compliance costs reduced, citizens spent their time using the self-service stations, often without success.

In Brazil, many citizens wanted to use the myINSS digital intermediary as it could bridge burdens of time and traveling costs. However, issues with authentication and system crashes often made the digital intermediary virtually unavailable, failing in bridging such compliance costs. One Brazilian citizen stated *"I would rather not have to come to the unit. I wish I had done it through the app. Call center? Forget it. Wait for hours and, when they pick up, they just schedule you to go to the unit."* The lack of functionality of the digital intermediary posed extra burdens in terms of time and effort as citizens could not comply with regulations and make their applications reach INSS case-handling officials: *"I contacted INSS many times, the (IT) system is always down. Always via telephone… You cannot go to the office without an appointment. I don't use the app, it's also usually down… They dismissed my request on the phone, before setting the date, it was rejected before the service. I must bring documents to prove that I have no income."* When the system worked, it still could impose compliance costs on the citizens as their applications were rejected before they could comprehend the system's logics. One citizen stated: *"The ideal would be to schedule the appointment through the Internet. But the appointment I made was straightaway rejected when I made it on my own, automatically, without any explanation."*

The Role of Personal Intermediaries in Mitigating/Reducing Compliance Costs
When citizens struggled to grasp the rules of the system and when frontline workers were perceived as unhelpful, citizens in NAV often turned to friends, family or other citizens to get their businesses conducted. One frontline worker at a NAV office stated *"We assist in the best way we can. There are many citizens that require our help, and the time will not be enough to help everyone. We are not supposed to help citizens fill out service applications, that is not our role, but still, we do it to help."*

To bridge compliance costs in Brazil, citizens also turned to their private network, but more often they relied on paid private professional brokers, the Despachantes, who sometimes were found visiting the observed INSS offices accompanying or on behalf of citizens to manage administrative tasks or negotiate with INSS officials. The Brazilian Despachantes are self-employed bureaucracy experts, who charge citizens for assisting or representing them in public service encounters [7]. They ease citizens from burdens of having to travel to INSS offices, as many citizens lived far away, had other obligations to attend, or struggled to afford transportation. One citizen that could not afford such brokers explained: *"It would be better not to go to the unit, as there is too much expense. I am unemployed, and I must pay for the bus tickets. My sister accompanied me."* The Despachantes are often acquainted with regulations, eligibility criteria and what the citizens need to successfully enroll in the service procedure. However, their services also impose a financial compliance cost for the hiring citizens.

In both NAV and INSS, frontline workers had no formal role to assist citizens in the self-service as by design, they were not supposed to help citizens in self-services. In NAV, they often helped citizens conduct self-service within a restricted time frame as they tried to manage informal guidance and assist many citizens with different tasks.

4.3 Psychological Costs

In this section, we discuss how citizens in NAV and INSS experienced psychological costs of mental strain and autonomy loss in their digital government encounters and how these costs are imposed/worsened by a) digital intermediaries and mitigated/reduced by b) personal intermediaries in self-service co-production.

The Role of Digital Intermediaries in Imposing/Worsening Psychological Cost

In the Norwegian case, many citizens experienced psychological costs due to the lack of autonomy. Those who were not comfortable with digital tools and preferred paper applications felt stressed and unsafe when they were pushed into a digital application process by NAV's "digital-first" approach. This loss of autonomy resulted in psychological costs such as feeling stressed, unsafe, and uncomfortable without the support and guidance of a frontline worker.

Many citizens that used the self-services at INSS had problems with authentication (login, mainly concerning their access password). As one citizen stated *"The system does not work most of the time. To book today, I called 20 times. The (digital) system is frequently down...I tried a few times to use the app, but it has always some error. Password locked, invalid user.... To get a new password you must go to the INSS office."* At one point, there was a long queue for self-services at the INSS office that made people feel stressed. At the waiting hall, one person was crying because they had missed their appointment as non-booked people have to justify their presence in the office. The fact that myINSS, where citizens can book their appointments, seldomly worked as they expected created a lot of stress and frustration. Those are examples of how the digital intermediary can impose psychological costs on citizens when technology is not working for them.

The Role of Personal Intermediaries in Mitigating/Reducing Psychological Cost

In both NAV and INSS citizens came to the offices because they face difficulties using myNAV and myINSS. Often, they were directed to partly guided self-service space by frontline workers. Citizens that were not accompanied by a private intermediary (family/friends) were dependent on the help from the outsourced employee in INSS and frontline workers in NAV. When they were not able to assist the citizens due to time constrains, stress and lack of knowledge, this posed stress on citizens. Some citizens experienced bad treatment from the frontline workers that increased psychological costs. One citizen at INSS stated: *"I always feel that they treat me with disregard. It feels like I'm asking for a favor. Begging for help."* This shows that the attitude of the personal intermediary will impact the mitigation of psychological cost and stresses the importance of human sympathy and service mindset. A Norwegian citizen expressed: *"I just wanted to talk to a person that can listen to me and provide me with some guidance. That's why I come today with my parents, I need the personal support"*. Often the intermediary does not need to be active, just being there and giving confirmation will ease psychological cost, as one official at INSS mentioned: *"Guided self-service does the trick: often people just want someone to confirm that what they're doing is right"*. The support of a Despachante in Brazil can provide that sympathy to ease the psychological costs, making citizens feel safe to rely on the intermediary. A citizen at INSS stated *"If I started the (former) process with a lawyer (often lawyers act as Despachantes and Despachantes present themselves as lawyers) I would have a better chance. (That time) my first channel was the lawyer. An assistant of the lawyer brought me here, but he stays outside."*

5 Discussion and Conclusion

Our study finds that citizens who visit the office often experience challenges with digital self-services that is in line with previous studies [8]. These citizens are more reliant on personal intermediaries to bridge the burdens often laid by digital intermediators. Out of this our findings have two main contributions as we answer our research question *"What is the role of intermediaries for mitigating the burdens citizens experience when co-producing digital self-services?"*

First, we observe that many citizens that come to the office can be considered digitally disadvantaged as they are not able to use digital self-service remotely [13] and face administrative burdens when utilizing digital intermediaries in the co-production process. This is the case in both the Brazilian and the Norwegian self-service context, supporting the validity of the findings across diverse contexts [27]. While digital intermediation aims to enhance collaboration between citizens and the government [39] in ways that enriches co-production [8], our findings reveal instances where digital intermediaries in effect complicate the interaction for citizens. This includes information presented on webpages being written in a bureaucratic language, and mismatching information—such as the government's use of different names for services on digital platforms—cause comprehension hardships and confusion among citizens. Consequently, digital self-services can lead to stress and a sense of losing autonomy for citizens, in fact preventing co-production from being realized. In the case of Brazil, the adoption of digital self-services could effectively increase financial burdens for citizens, as they are forced to rely on Despachantes' paid services to access these services [33]. Those represent, in fact, additional compliance costs, worsening citizens' financial struggles and scarcity in an already strained situation. Besides, the Brazilian digital intermediary often failed to bridge citizens' burdens through digital self-service procedures, because of technological and authentication (login) issues, making it virtually unreliable. For citizens in Norway, the digital systems worked and were deemed as reliable but citizens' ability to use them often caused constraints and additional burdens for citizens.

Second, our findings also highlight that personal intermediation helps ease burdens in co-production. We found that the different categories of personal intermediaries, both 'unpaid' and 'paid' intermediaries can reduce citizens' experiences of administrative burden. In both Brazil and Norway, intermediaries advised citizens on complying with rules and regulations. They assisted citizens in understanding requirements and provided guidance throughout the self-service procedures. Additionally, family members and friends in both countries helped citizens with asking the right questions, interpreting eligibility criteria, and interacting with self-service stations at the offices. In Norway, personal intermediaries were often frontline workers in NAV, although they were not officially responsible for assisting citizens in the self-service procedure. They managed citizens' digital service applications on their behalf through their logged-in profiles at myNAV. While the intermediaries helped alleviate burdens in this situation, citizens still struggled to understand the procedure. This demonstrates the importance of intermediaries for citizens to reduce their burdens in the present. In Brazil, Despachantes often took on tasks from citizens, easing their compliance costs. However, citizens without intermediaries were left without support, emphasizing the significance of intermediaries in reducing citizens' experiences of administrative burdens.

In our cases an emerging pattern is that digital intermediaries typically create administrative burdens while personal intermediaries in most cases help reduce barriers. This makes the role intermediaries play in mitigating administrative burdens in digital coproduction complex, as intermediates have the potential to both reduce and impose burdens on citizens. Our understanding of digital self-service technology is that it works as a digital intermediary by mediating the interaction between citizens and government agencies. The emphasis on digital intermediaries has previously been explored from a more positive and nuanced perspective [39] that our findings challenge. We do not deny that intermediaries have potential to support citizens in their service interactions as selfservice may ease their experience of burdens when reducing compliance costs of having to physically go to an office. This is often an argument in favor of digital self-services as they enable service access at any hours from the comfort of citizens' preferred location [8, 9]. Besides, the psychological cost of shame in relation to social security services can be reduced as the citizen can be anonymous when applying from home [9]. In this way, digital intermediaries can create a space for those citizens that manage to co-produce their digital self-services, enabling learning and knowledge building as they administrate their own services. To account for such connections is a limitation of this study as we do not focus on citizens that successfully interact with the digital self-services remotely. For those citizens the digital intermediary can have different meanings and may provide streamlining access to digital services that reduce burdens.

In our case studies we see that intermediaries can take different forms in different contexts and be positioned both inside and outside the public service organizations. This answer to the call to understand third party actors role in administrative burden context [21]. This is also interesting because intermediaries have mostly been studied as separate units from the service organization [13, 38]. This makes such inside/outside relation and the shifting intermediary roles, interesting to explore in further research.

Finally, our examples also demonstrate that citizens who encounter significant administrative costs when utilizing digital self-services do not necessarily achieve the expected empowerment promised by self-service co-production [30, 31]. This study supports previous findings that digital self-services are changing the roles of citizens and frontline workers, as described in existing literature [8, 15]. However, the cases from Norway and Brazil demonstrate a new aspect of this phenomenon, in which intermediaries become co-producers of self-services. This aspect challenges earlier literature emphasizing the importance of citizen involvement in self-service processes and co-production as inseparable from the service [28]. This perspective is intriguing, as it suggests a *hybridity* between digital and personal Intermediaries, through a triangulation effect, which deserves further research. The level of activity and ownership the intermediating agents are given/take and how this impact citizens in the self-service co-production should also be further explored. Finally, our study has not evaluated intermediation from the citizen's perspective, where we have seen how they often depend on personal intermediation for success. If citizens perceive this as good or bad should be object of furthers studies.

Disclosure of Interests. The authors have no competing interests to declare that are relevant to the content of this article.

References

1. Breit, E., Egeland, C., Løberg, I.B., Røhnebæk, M.T.: Digital coping: how frontline workers cope with digital service encounters. Soc. Policy Admin. **55**(5), 833–847 (2021)
2. Busch, P.A., Henriksen, H.Z.: Digital discretion: a systematic literature review of ICT and street-level discretion. Inf. Polity, **23**(1), 3–28 (1999). Author, F., Author, S., Author, T.: Book title. 2nd edn. Publisher, Location
3. Alford, J.: Engaging Public Sector Clients: From Service-Delivery to Co-Production. Springer, Cham (2009)
4. Hilton, T., Hughes, T.: Co-production and self-service: the application of Service-Dominant Logic. J. Mark. Manage. **29**(7–8), 861–881 (2013)
5. Osborne, S.P., Radnor, Z., Strokosch, K.: Co-production and the co-creation of value in public services: a suitable case for treatment? Public Manage. Rev. **18**(5), 639–653 (2016)
6. Giest, S., Samuels, A.: Administrative burden in digital public service delivery: the social infrastructure of library programs for e-inclusion. Rev. Policy Res. **40**(5), 626–645 (2023)
7. Haug, N.: Actor roles in co-production—Introducing intermediaries: Findings from a systematic literature review. Public Admin. (2023)
8. Rydén, H.H., Hofmann, S., Verne, G.: The self-serving citizen as a co-producer in the digital public service delivery. In: Lindgren, I., et al. (eds.) EGOV 2023. LNCS, vol. 14130, pp. 48–63. Springer, Cham (2023). https://doi.org/10.1007/978-3-031-41138-0_4
9. Heggertveit, I., Lindgren, I., Madsen, C.Ø., Hofmann, S.: Administrative burden in digital self-service: an empirical study about citizens in need of financial assistance. In: Krimmer, R., et al. (eds.) EPart 2022. LNCS, vol. 13392, pp. 173–187. Springer, Cham (2022). https://doi.org/10.1007/978-3-031-23213-8_11
10. Janssen, M., Klievink, B.: The role of intermediaries in multi-channel service delivery strategies. Int. J. Electron. Gov. Res. (IJEGR) **5**(3), 36–46 (2009)
11. Perry, B., Smit, W.: Co-producing city-regional intelligence: strategies of intermediation, tactics of unsettling. Reg. Stud. **57**(4), 685–697 (2023)
12. Madsen, C.Ø., Kræmmergaard, P.: Warm experts in the age of mandatory e-government: Interaction among Danish single parents regarding online application for public benefits. Electron. J. E-Gov. **14**(1), 87 (2016)
13. Alshallaqi, M., Al-Mamary, Y.H.: Paradoxical digital inclusion: the mixed blessing of street-level intermediaries in reducing administrative burden. Gov. Inf. Q. **41**(1), 101913 (2024)
14. Herd, P., Moynihan, D.P.: Administrative Burden: Policymaking by Other Means. Russell Sage Foundation, New York (2019)
15. Madsen, C.Ø., Lindgren, I., Melin, U.: The accidental caseworker–how digital self-service influences citizens' administrative burden. Gov. Inf. Q. **39**(1), 101653 (2021)
16. Bartlett, L., Vavrus, F.: Comparative case studies: an innovative approach. Nordic J. Comp. Int. Educ. (NJCIE), **1**(1) (2017)
17. de Andrade, L.H.A.: Avaliação de desempenho na experiência de gestão colaborativa do INSS no Seguro Defeso. Planejamento e Políticas Públicas **56**(October 2020), 221–250 (2020)
18. Paula Regina Wenceslau Lloyd and Jucilaine Neves Sousa Wivaldo. 2019. Meu INSS: Inclusão ou Exclusão? Assistência Social em Foco 1, 1 (November 2019), 11
19. Hansen, H.T., Lundberg, K., Syltevik, L.J.: Digitalization, street-level bureaucracy and welfare users' experiences. Soc. Policy Admin. **52**(1), 67–90 (2018)
20. Nye Arbeids- og Velferdsetaten Nav. 2021. Nav's Horizon Scan 2021 - Report 1. Directorate of Labour and Welfare, Oslo. https://www.Nav.no/no/Nav-og-samfunn/kunnskap/analyser-fra-Nav/Nav-rapportserie/Nav-rapporter/omverdensanalyse-2021
21. Halling, A., Baekgaard, M.: Administrative burden in citizen-state interactions: a systematic literature review. J. Public Admin. Res. Theory **34**(2), 180–195 (2024)

22. United Nations UN. 2024. EGOVKB - United Nations E-Government Knowledgebase. https://publicadministration.un.org/egovkb/en-us/ (2024). Accessed 21 Feb 2024
23. Pettersen, P.A.: Welfare state legitimacy: Ranking, rating, paying: the popularity and support for norwegian welfare programmes in the mid 1990s. Scand. Polit. Stud. **24**(1), 27–49 (2001)
24. Schwarzer, H., Querino, A.C.: Benefícios Sociais e Pobreza: Programas Não Contributivos da Seguridade Social Brasileira. Textos para Discussão **929**(December 2002), 55 2002
25. Seawright, J., Gerring, J.: Case selection techniques in case study research: a menu of qualitative and quantitative options. Polit. Res. Q. **61**(2), 294–308 (2008)
26. Hsieh, H.-F., Shannon, S.E.: Three Approaches to Qualitative Content Analysis. Qual. Health Res. **15**(9), 1277–1288 (2005)
27. Eisenhardt, K.M., Graebner, M.E.: Theory building from cases: opportunities and challenges. Acad. Manag. J. **50**(1), 25–32 (2007)
28. Osborne, S.P., Strokosch, K.: It takes Two to Tango? Understanding the C o-production of public services by integrating the services management and public administration perspectives. Br. J. Manage. **24**, S31-S47 (2013)
29. Pestoff, V.: Co-production and third sector social services in Europe: some concepts and evidence. Voluntas: Int. J. Voluntary Nonprofit Organ. **23**, 1102–1118 (2012)
30. Eriksson, K., Vogt, H.: On self-service democracy: configurations of individualizing governance and self-directed citizenship. Eur. J. Soc. Theory, **16**(2), 153–173 (2016). Author, F., Author, S.: Title of a proceedings paper. In: Editor, F., Editor, S. (eds.) CONFERENCE 2016, LNCS, vol. 9999, pp. 1–13. Springer, Heidelberg
31. Harrison, T., Waite, K.: Impact of co-production on consumer perception of empowerment. Serv. Ind. J. **35**(10), 502–520 (2015)
32. Larsson, K.K., Skjølsvik, T.: Making sense of the digital co-production of welfare services: using digital technology to simplify or tailor the co-production of services. Public Manag. Rev. **25**(6), 1169–1186 (2023)
33. Meijer, A.: Coproduction as a structural transformation of the public sector. Int. J. Public Sect. Manag. **29**(6), 596–611 (2016)
34. Burden, B.C., Canon, D.T., Mayer, K.R., Moynihan, D.P.: The effect of administrative burden on bureaucratic perception of policies: evidence from election administration. Public Admin. Rev. **72**(5), 741–751 (2012)
35. Cordella, A., Tempini, N.: E-government and organizational change: reappraising the role of ICT and bureaucracy in public service delivery. Gov. Inf. Q. **32**(3), 279–286 (2015)
36. Bailey, J.P., Bakos, Y.: An exploratory study of the emerging role of electronic intermediaries. Int. J. Electron. Commer. **1**(3), 7–20 (1997)
37. Sein, M.K., Furuholt, B.: Intermediaries: bridges across the digital divide. Inf. Technol. Dev. **18**(4), 332–344 (2012)
38. Sorrentino, M., Niehaves, B.: Intermediaries in E-inclusion: A literature review. In: Proceedings of the 43rd Hawaii International Conference on System Sciences, January 5 – 8, 2010, Kauai, HI (2010)
39. Josefsson, U., Ranerup, A.: Consumerism revisited: the emergent roles of new electronic intermediaries between citizens and the public sector. Inf. Polity **8**(3–4), 167–180 (2003)

Getting Rid of It: An Unlearning Perspective on Digital Government Competences

Michael Koddebusch[1](✉)[iD], Bettina Distel[1][iD], Marco Di Maria[2][iD], Paul Brützke[1][iD], and Jörg Becker[1][iD]

[1] University of Münster, Münster, Germany
{michael.koddebusch,bettina.distel}@ercis.uni-muenster.de
[2] University of Hildesheim, Hildesheim, Germany
marco.dimaria@uni-hildesheim.de

Abstract. Digital Government Competences are vital for digital transformation in public administration. However, research has mostly focused on the acquisition of new competences, neglecting prevailing ones that contradict transformation efforts. We address this shortcoming by adopting an unlearning perspective, which depicts the process of discarding obsolete competences that hinder rather than foster change. Consequently, we present the results of a focus group study with 34 public servants and subsequent in-depth qualitative analysis. We apply the lens of an established competence framework that defines competence dimensions as individual and organizational as well as visible (observable behavior) and hidden (cognitive mechanisms). The results identify both individual and organizational competences in 24 categories that must be unlearned to unlock the potential of digital transformation in public administration without legacy burdens. We contribute to digital government research by complementing the competence discourse with an unlearning perspective and offer practitioners a new angle on continuous education.

Keywords: Unlearning · Digital Government · Competences · Digital Transformation · Focus Groups

1 Introduction

When changes occur in an organization's environment, the organization needs to adapt to maintain its function and be able to further create value. One such change, posing a particular challenge to public administrations, is the digital transformation [46]. Scholars from various disciplines devoted their work to describing and conceptualizing digital transformation in public administration [21,35], what hinders and drives digital transformation in the public sector [45], and how stakeholders adapt to changes related to digital transformation. Others focused on the societal consequences of a digitalized public sector [17]. These studies, however, primarily focus on the organizational level of change.

Public administrations, just as organizations in general, are constructs essentially consisting of individuals working together to fulfill a shared purpose [4]. In accordance with earlier works [8], we argue that digital transformation starts at the individual level and transcends up to the organizational level. Thus, digital transformation depends on the employees' personality and ability to support the change. Here, scholars repeatedly discussed the importance of acquiring the competences necessary to master the digital transformation in the public sector, i.e., *digital government competences*. They have not only provided comprehensive lists of specific competences [13,22, e.g.,], but also suggested measures for imparting those [38, e.g.,]. To ground our research, we base our understanding of competences on the seminal work by Bloom et al. [5] and understand competence as the combination of knowledge and skills. More recent research indicates that additionally one's personality traits and motives also form a component of competences [13, e.g.,], because they act as conscious and unconscious drivers of behavior [9]. Integrating prior research, we therefore define *digital government competences* as *the combination of a public servant's skills, knowledge, self-concept, and motives to act in a digitally transforming public sector.*

The acquisition of new digital competences has received considerable attention in digital government research (and beyond). However, for public organizations to master digital transformation, it is not only crucial to acquire new competences but also to discard those contradicting the efforts of digital government transformation. The process of discarding knowledge, skills, and routines that contradict the change processes, and rather hinders than fosters change, has been termed *organizational unlearning* [24]. Although research on unlearning within digital government is still in its early stages, initial studies depict the value of exploring unlearning in this context. For instance, Pan et al. [39] explored the role of unlearning entrenched beliefs in the turnaround of an e-government project. Others [13,19, e.g.,] emphasize the necessity for public sector organizations to discard outdated competences to keep pace with technological advances. Danneels and Viaene [11] discuss the difficulties public servants face in breaking long-established mental and behavioral patterns that hinder transformation efforts. Similarly, Matsuo [31] underlines the importance of critical reflection as a tool for unlearning obsolete practices among municipal employees. Prabowo et al. [42] investigate how unlearning can help reduce corruption in public institutions to enhance governance.

Despite these works' results are intriguing, unlearning in the context of digital government has been underrepresented, overshadowed by a predominant focus on imparting new competences. A comprehensive framework identifying which specific competences need to be unlearned is missing. Problematic to this one-sidedness are the conflicts and tensions arising from the prevalence of established routines, skills, and behaviors on the one side and attributes of the newly introduced systems on the other side. Tensions arise primarily from the meaning individuals attach to established routines and behaviors, which is certainty and security [8], but also from subconscious habits that shape responses to new systems [1,41]. Solely focusing on learning new competences and routines likely results in failed organizational transformation, wherefore we complement the emphasis on learning competences with a perspective of unlearning obsolete competences by answering the following research question:

Which competences must be unlearned in public administration on an individual and organizational level to facilitate digital transformation?

2 Research Background

The interplay between individual and organizational competences is a dynamic process where employees' individual competences and use of technology interact to form organizational core competence [27]. Expanding on Lahti's [27] ideas, Chen and Chang [9] propose a framework that integrates both levels. Accordingly, individual competence comprises skills (abilities to demonstrate behaviors to attain goals), knowledge (usable or practical information), self-concept (how people behave and feel in a social context), and motives & traits (what people want and what their personality is). Organizational core competence entails strategic skills and knowledge (portfolio of employees' individual skills and knowledge and subject to strategic development), organizational image (the image each member has of the organization), and strategic intents (shared sense of obligation to external challenges among members of the organization) [9]. The authors further differentiate between the "visible" and "hidden" dimension of competence. The visible dimension (skills/knowledge/strategic skills/strategic knowledge) is observable, directly related to performance, and developed according to strategic orientation. The hidden dimension (self-concept/motives & traits/organizational image/strategic intents) depicts diffuse role- and context-dependent cognitive mechanisms in interactive settings [9]. Due to its comprehensiveness, we chose this as the analytical framework for our work. While individual competences are dynamic, organizational ones tend to be stable and underpin long-term value creation. This poses the challenge of preserving core competences while adapting to environmental shifts. Formerly useful competences may turn into rigidities [29]. Therefore, public administrations must navigate between conserving existing competences and pursuing innovation [7].

To this end, the process of unlearning becomes instrumental in helping organizations realign their competences with emerging needs. This process, addressing both individual and organizational levels, contemplates the dynamism between current and required competences [33]. Initiated by individuals, unlearning progressively influences the whole organization [15,28]. Individually, unlearning involves recognizing, examining, and potentially letting go of obsolete competences, thereby challenging the tendency to cling to established practices [34]. At the organizational level, it entails moving beyond prevailing logics [3] and antiquated routines [16] that have inhibitive effects. Reflecting on Lahti's [27] division of competences (visible/hidden), Klammer and Gueldenberg [25] distinguish between shallow and deep knowledge structures, whereby unlearning starts with the critical act of sensing: the recognition and reassessment of what competences might be obsolete [18]. This refined perspective highlights the need for a deliberate and reflective approach to unlearning, enabling organizations to navigate and adapt to the continuously evolving landscape.

In government sectors, where stability often prevails, legislative or technological changes can rapidly question the utility of existing competences [28,43]. Thus, fostering unlearning becomes crucial for public organizations to adapt and remain effective [10,30]. To this end, employee competences are widely recognized as one of the key pillars for digitally transforming structures and processes [36]. Similarly, Distel et al. [13] underscore the critical role of digital competences and the challenges posed by a rapidly evolving competence landscape, pointing to a significant knowledge gap. This indicates the need to identify, reflect upon, and discard obsolete competences that counteract the future-orientation of public administration [40].

3 Research Method

To answer our research question, we applied a qualitative approach using focus group discussions (further called *focus groups*). Focus groups have long been an established research method in many scientific disciplines [23,37] and have soon found their way into the field of information systems research, too [44,47]. Focus groups bring together individuals to discuss a given topic. The researchers act as moderators and observers who provide the topic and questions for the discussion. The value of data lies within the exchange of opinions between the participants, varying perceptions of the phenomena discussed, and gaining a holistic understanding of a topic. We decided on this research method as *unlearning* is a rather theoretical concept, which we deemed to be challenging to grasp for individuals in 1-to-1 interviews. Moreover, we planned to gather insights on unlearning in public organizations from individuals exchanging opinions and discussing experiences. Our sampling strategy followed a purposeful and self-selection paradigm [6]. We reached out to municipal administrations, outlining the planned research endeavor and criteria for participants (different departments, genders, ages, years of experience, digital savviness, and hierarchy levels). This request was then forwarded to the workforces. We conducted five focus groups with six to eight participants, resulting in 106 min of recording. The focus groups took place in October and November 2023 and included 34 participants from nine German municipal administrations. Table 1 shows an aggregated distribution of the roles and responsibilities of the participants across the municipal administrations. Exemplary questions for the focus groups were:

1. Is there anything you learned in your education that you have had to use less and less over the years and is becoming increasingly obsolete?
2. What specific working methods/routines/skills do you think will (have to) be less common in the future?
3. Are there developments in your working environment that influence "unlearning"? What are these developments or could they be?

Table 1. Focus Group Discussants' Roles and Responsibility Areas

	IT	Digitalization	Organisation	Personnel	Central Services	Construction
Leadership	3	2	1	2	1	2
General workforce	2	4	4	4	6	3

The recordings were transcribed automatically using software, followed by manual revision. The transcripts were subject to qualitative content analysis as per Mayring [32], applying an inductive-deductive coding strategy. The deductive part of the coding aimed at sorting the data into the competence framework by Chen and Chang [9] (cf. Chapter 2). The results were enriched by inductively emerging concepts, summarized under the headline *general insights* (cf. Chapter 4). After establishing the codes, they were further grouped in two subsequent iterations (cf. Fig. 1). First, the concepts that emerged from the deductive coding were categorized according to their thematic focus. Afterwards, the categories were grouped into themes to describe more abstract phenomena.

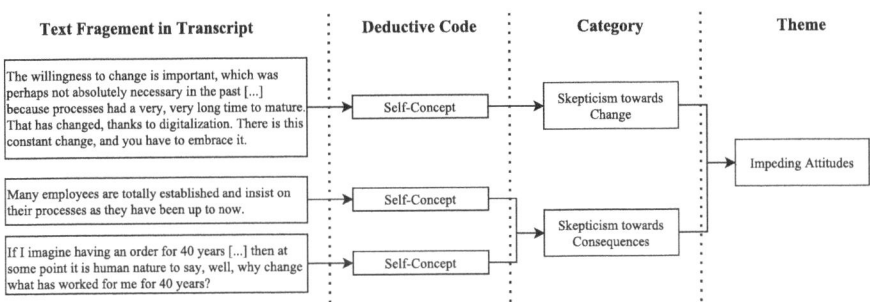

Fig. 1. Exemplary Code Categorization Process

4 Results

This chapter illustrates the insights of our research, defining *what* needs to be unlearned in public administration to facilitate digital transformation. The results are summarized in Table 2. The former two columns depict the competence dimensions as per Chen and Chang [9]; the latter two columns allocate unlearning themes and categories as emerged from the analysis to the respective dimensions.

4.1 Preliminary Analysis

Based on the definition of Chen and Chang [9], both individual competence and organizational core competence comprise "visible" and "hidden" dimensions (cf.

chapter 2). Across all five focus groups, the participants mainly discussed the hidden dimension of competence and some participants articulated struggling with the concept of unlearning. Statements like *"For me, it has simply always been a process and [...] cultural change [...] continues to evolve, and that leads to me changing the way I see things today compared to how I saw them in the past. And I consciously don't actively forget or discard anything but rather continue to develop."* were met with broad approval by most participants. Discussions concluded quickly when discussing (visible) skills and knowledge. The necessity of unlearning anything was not evident, even if competences were deemed no longer useful. In contrast, more diffuse (hidden) subjects were discussed extensively. This might correlate with their understanding of the concept of competence. When facing the question of what *being competent* is, answers predominantly revolved around *knowing-*, *being able to do-*, or even *conveying* something: *"I feel competent when I understand what I'm talking about. When I have penetrated a topic, can understand and categorize the contributions of others and perhaps even assess whether someone else understands what they're talking about.* The "hidden" dimension of competence, however, was seldom mentioned.

Furthermore, participants switched between statements regarding individuals and the overall workforce. When treating one's own unlearning, the participants primarily referred to actual habits and skills: *"So what I consciously try to do, I don't actually use a pen and pencil in meetings anymore.* More general considerations concerning the overall workforce partly revolved around skills and knowledge but mainly considered parts of the hidden competence dimensions, which shows that competence is perceived as a complex mesh of interrelated, overlapping, and non-exclusive concepts: *"If they [leadership] live the silos, then the rest lives the silo."*

4.2 Unlearning on an Individual Level

Skills: Most skills are viewed as evolving rather than becoming obsolete. In this light, the following *skills* are not necessarily *skills* to be unlearned in the conventional meaning but refer to patterns and practices that impede the development of skills. These, we subsumed under the two themes *restrictive thought patterns*, and *organizational incapacities*.

Starting with restrictive thought patterns, *paralyzing indifference* refers to the unrestricted prioritization of administrative instructions. This prioritization is a long-learned habit of not risking non-compliance with procedures and rules. However, public servants today have creative leeway towards their processes and actions: *"It's important to have learned the basics of administrative action and I'm not saying that I don't longer need it. But in my current role I need less knowledge and more questioning."* Paralyzing indifference towards these instructions must be unlearned to deal with competing priorities.

Passive acceptance describes the individual acceptance of the structures one works in instead of questioning these. Working environments are often condoned without critically being reflected, neglecting the potential for improvement. *"I think there's a lack of flexibility and [...] personal responsibility [...]. Thinking for*

Table 2. Digital Government Competences to be Unlearned

Level	Code	Theme	Category
Individual Competence	Skill	Restrictive Thought Patterns	Paralyzing Indifference
			Passive Acceptance
			Decision-Making Diffusion
		Organizational Incapacities	Processual Apathy
			Physical-Only Self-Organization
	Knowledge	Processual Unawareness	
	Self-Concept	Impeding Attitudes	Scepticism towards Change
			Scepticism towards Consequences
		Organizational Misconceptions	Misconceptions of the Organization
			Misconceptions of the own Place
			Misconceptions of the own Purpose
	Motives and Traits	Mistrust in Technology	
Organizational Core Competence	Strategic Skills	-	
	Strategic Knowledge	Digital Transformation Misconceptions	Scope Misconcpetions
			Consequence Misconceptions
	Organizational Image	Distorted Organization Image	
		Canonization of Administrative Instructions	
		Misconception of Roles and Responsibilities	Leadership Misconceptions
			General Workforce Misconceptions
	Strategic Intents	Lack of shared Purpose	
		Structural Transformation Barriers	Concentration on few Decision Makers
			Organizational Change Lethargy
		Strategic Gaps	Performance Management Gaps
			Knowledge Management Gaps

yourself about what you're actually doing: does it really make sense?" Unlearning this passive acceptance is likely to precede critical thinking.

Decision-making diffusion depicts the unwillingness to make decisions based on the perceived risk of being liable for errors. The discussants talked about individuals' likeliness of deferring decisions to higher hierarchy levels, resulting in long and tedious decision-making processes: *"The problem is that everything takes so long [...]. Leadership doesn't have the decisiveness that it perhaps should have."* Unlearning decision-making diffusion on an individual level might facilitate the acquiring of decision-making skills.

In terms of *organizational incapacities*, the participants discussed individuals' *processual apathy*. Apathy in this context is the perceived inability to change something about the processes one partakes in. *"It's always this 'check responsibility first [...]. But processes and responsibilities have to be a bit more flexible to solve things."* Process management is an important competence but must be preceded by unlearning processual apathy.

Physical-only self-organization practices were mentioned several times. Individuals retain their self-organization habits from non-digital times, i.e., processing physical work and neglecting work in electronic file systems. *"Back then, you had your folders with papers inside, and now you must organize yourself digitally. [...] It's this remembering that you have to look digitally and not just at what's on your desk."*

Knowledge: In the context of unlearning, we understand *knowledge* as knowledge-related assets that hinder the strategic pursuit of digital transformation. The only asset the participants came up with on an individual level was *processual unawareness*, which describes not having a clear end-to-end view of the processes one partakes in: *"What comes to my mind [in terms of unlearning] is what we [...] call service by the book. When you get a task, you work through it, pass it on and don't care what happens afterwards."* This unawareness leads to the inability to assess the consequences of actions and its unlearning is expected to bear the potential to enhance the overall process quality.

Self-concept: Describing the individual behavior and feelings in social settings, the two major themes to be unlearned in this dimension are *impeding attitudes* and *organizational misconceptions*. A general *skepticism towards change* was attested to many public servants, resulting in an unwilling and hesitant adoption of new processes and technologies. Examples were given in the context of digital processing times: *"The willingness to change is important, which was perhaps not absolutely necessary in the past [...] because processes had a very, very long time to mature. That has changed [...]."* These attitudes were discussed in light of the frequent absence of willingness to change and the fact that being open and enthusiastic about change is rare. Moreover, the *skepticism towards consequences*, which individuals face in terms of work routines and practices, is often negatively connoted. *"Many employees [...] insist on their processes as they have been up to now."* Unlearning these impeding attitudes was considered a prerequisite for facilitating future competence acquisition.

Organizational misconceptions revolve around the individual perception of the organization and one's place and purpose therein. Regarding *perception of the organization*, public servants primarily focus on their jobs and departments, negating a shared organizational purpose. Instead of cooperating to achieve better outcomes, individuals perceive competition between departments, which hinders organization-wide change processes, such as digital transformation.

Concerning *perception of one's place in the organization*, we observed that individuals view themselves more as workers receiving and carrying out tasks than as initiative-taking designers of the organization. This is characterized by, for example, the fear of being punished for mistakes. *"One time, something really went wrong, and somebody faced serious repercussions. Making errors can have consequences, and that generates fear."* This fear enforces *playing by the rules* and, thus, can stoke solution-orientation.

The *perception of one's purpose in the organization* points to a self-assessment of one's contribution, i.e., task orientation instead of service orientation. Many individuals work self-referentially to do their work well but forget that it is part of larger service delivery, leaving quality measures such as efficiency or resource effectiveness out of account. *"Individual task fulfillment is the absolute priority [...]."* However, limiting one's perception of responsibilities again enforces problems like silo-thinking and lack of shared purpose. Moreover, this discussion led to the role of leaders in public organizations. According to the participants, leaders often avoid conflict and aim to be likable. This impacts their job of addressing questions of individual performance or initiating organizational changes, which hinders change projects on a broader scale. *"They simply shy away from conflict and confrontation [...]. Some people only get the minimum points in the performance appraisal, but very few of their leaders dare to talk to them honestly about it."* Leaders also exemplify that established structures, orders, and instructions should be prioritized over creative and solution-oriented thinking and thus prevent their subordinates from coming up with innovative solutions to issues they recognize in the organization. In line with that, leaders act as omniscient individuals rather than benevolent managers who focus on developing their subordinates' competences.

Motives and Traits: The only concept sorted in this dimension is a still prevalent *mistrust in technology*. This mistrust is not necessarily understood as a general unwillingness to work with information systems but rather deeply anchored beliefs that physical artifacts are preferable to digital ones. Examples were given in the context of automated payment verification or round-robin folders. *"In finance, particularly in managing invoice workflows, there traditionally have been multiple stages requiring manual verification. However, with digitalization, invoices are now automatically processed and integrated into systems, eliminating the need for manual checks. Can and should we trust the system?"* Unlearning this mistrust can lead to a broader openness for new digital solutions.

4.3 Unlearning on an Organizational Level

No text was coded as **Strategic Skills**. This aligns with the discussants' challenges to depict unnecessary skills, as it is even harder to abstract from the individual to the organizational level.

Strategic Knowledge: All discussed knowledge-assets can be subsumed under *digital transformation misconceptions*.

As for the *scope of digital transformation*, the discussants highlighted that digital transformation is often solely understood as the usage of new technologies: *"Everyone wants an iPad, so I say, 'Sure here's an iPad!'. And now? Well, I don't have an answer!"*. It has yet to seep through that digital transformation is entirely comprised of more than just new technology; it affects individual and organizational work routines, habits, and values. Unlearning this narrow understanding of digital transformation is essential for comprehending the *consequences of digital transformation*. *Consequences* reflect the workforce's perceived impact of digital transformation on collaboration and cooperation within the workplace. One participant described an experience of introducing instant messaging: *"Messaging is still viewed very critically in public administration because the content is not being archived. No one can track the answer to a question in weeks from now."* Although legal questions were accounted for, public servants were reluctant to use the new system due to liability reservations based on the perceived risk of usage. As digitalization-induced changes interfere with long-learned organizational practices, they are often met with skepticism.

Organizational Image: Representing an image that the overall staff shares, this section presents three themes: *distorted organizational image, canonization of administrative instructions,* and *misconception of roles and responsibilities.*

Distorted organizational image refers to a widespread conviction that working inside public administration is honorable and (especially young) people should be grateful. This point of view denies the public sector's challenges raised by the employee-favoring job market and consequential competition for talent: *"Public administration [in its self-perception] will have to change because otherwise [applicants] will just go elsewhere. Anyone who [...] doesn't feel comfortable will just go to another [public or private] organization."* New talent and skilled staff is imperative, so unlearning this distorted organizational image would likely result in public organizations improving their recruitment practices.

Canonization of administrative guidelines addresses the significance attributed to administrative guidelines from an organizational point of view: *"In public administration, everything is regulated by guidelines or instructions [...]. It is precisely regulated who can sign what, up to what value limit, what should the letter look like, and so on. There are still so many formalities[...]. It goes like this: you tell your boss, who tells his boss, who tells his boss, and then you get it back after eight months; if you're lucky."* These regulations are typically captured in administrative guidelines, which are often considered the highest priority. In light of digital transformation, however, public servants face competing priorities.

Misconceptions of roles and responsibilities point out the shared image of the two staff levels: *leadership* and *general workforce*.
Leaders are often not seen as role models but as self-sufficient individuals who are given the right to make decisions and give orders. However, the participants all agreed that leadership is crucial for the success of digital transformation, e.g. for making silo-thinking obsolete. *"I can still remember [...] a department lead really saying: 'with all due respect, what do I have to do with your department?' What he has got to do with my department? You've got to be kidding me!"* The awareness of leadership of their own importance in pursuing change projects has not yet reached full maturity. Instead, many leaders maintain the status quo without disruption, including how they handle performance issues. Most of them strive to be liked instead of initiating well-meant conflict conversations:

"From what I've noticed over many years, there's a lot of focus on feeling good and not enough on discipline." There are two facets to unlearning leadership misconceptions: first, leaders must unlearn not only to be decision-makers but to

lead by example. Second, the general workforce must unlearn perceiving leaders as bossy managers and formulate their requirements regarding good leadership.

General workforce misconceptions address how little competence is attributed to clerks: "*You're not actually allowed to do anything at clerk level. You can go to the toilet alone, but that's it in public administration.*" This quote repeats the issue that public servants "soley" carry out tasks they are told to, but elevates it to an organizational dimension. This also manifests in their often non-existent decision-making power: "*In our organization, we are not given the freedom to make any decisions at all.*" Here again, unlearning must happen on both sides: the general workforce must unlearn self-perceiving themselves as the ones carrying out tasks. Similarly, however, leaders must unlearn to restrict and distrust the general workforce and thus degrade them to pencil pushers.

Strategic Intents: Depicting a "shared sense of obligation to the external challenge" [9], we identified three overarching themes: the *lack of shared purpose*, *structural transformation barriers*, and *strategic gaps*.

The *lack of shared purpose* addresses single departments typically being stuck in closed-off thought patterns. There is little shared desire to achieve common goals: "*I was in the social welfare, finance, regulatory, and HR departments. At the end of the day, each department works independently [...]. It feels like everyone is unconsciously working against each other.*" Without this feeling of shared purpose, implementing digital transformation is challenging, as it requires the involvement of all individuals.

In terms of *structural transformation barriers* to be unlearned, we begin with *concentration on few decision makers*. Correlating with *decision-making diffusion*, decisions are made on the leadership level, and clerks rarely make autonomous decisions. On the one hand, this is due to decision-making latitude not being granted to lower-level public servants. However, even if this latitude is granted, public servants have long learned to defer decisions to the higher levels, which then again leads to higher turnaround times: "*Decision-making freedom is often restricted. Even when it is granted, it tends to go unused. Many of our department heads set a precedent by demonstrating cautious behavior. When facing questions, their typical response is, 'Yes, we need to ask for permission first, right?'*". That highlights that decision-making is not only personality-related but underlines a culture of dependency on higher-level approval.

Organizational change lethargy subsumes a prevailing belief for the non-necessity of change because practices have long fulfilled their purpose: "*[...] It's natural to question the need for change. Why alter something that has worked well for 40 years? However, applying these time-tested administrative methods to today's dynamic environment [...] inevitably leads to problems.*" The attitude was attributed to different contexts, such as organizational structures, work practices, hierarchical orders, and administrative principles. Interestingly, the participants referred to these beliefs not based on physical age but on professional lifetime spent in public administration: "*I don't mean the older generation in the sense of age, but people who have been in administration for a long time, who are already entrenched in their way of thinking [...]*".

Lastly, this section highlights *strategic gaps*, i.e., matters to be addressed by strategists. Unlearning strategic gaps can be understood as closing these gaps. *Performance management gaps* refer to an organizational understanding that fulfillment of tasks with maximum precision is the dominant quality criterion. How these tasks are fulfilled is secondary, i.e., whether something is done in an especially intelligent or efficient way. "*Individual task fulfillment is the absolute priority, and then there are the resulting costs [...]. However, the control bodies that we have are not really able to determine those at all.*" This understanding of performance should be unlearned because digital transformation largely depends on public servants' creativity and solution orientation.

Knowledge management gaps refer to the absence of structured knowledge management: "*There is so much tacit knowledge, especially among long-term employees. We must find a way to share this knowledge as recurring issues continually arise, consuming so much of our resources.*" This was raised in a discussion of the options around dealing with workload-related concerns of digital transformation. Unlearning these knowledge management gaps would prevent public servants wasting time solving already long-solved problems.

5 Discussion

5.1 Unlearning in Public Organizations

This article set out to explore competences that need to be unlearned in public organizations in order to facilitate digital transformation (Table 2). This article offers three central contributions: First, we identify a set of obsolete competences on both the individual and the organizational level. Our results indicate that unlearning in public organizations affects two components – specific skills and knowledge, and more intangible competence dimensions, mainly practices and thought patterns. Second, we find that the largest share of competences to be unlearned relates to the latter and is, thus, anchored in the hidden dimension of competences. Third, we find that frameworks of organizational unlearning are inherently analytical. In reality, public servants perceive the individual and organizational levels as well as the hidden and visible dimensions of digital government competences to be complex and strongly intertwined.

Regarding the first aspect, we find that many of the competences public servants deem obsolete are more work practices and thought patterns rather than "hard" skills. These less tangible competences, such as mistrust in technology, organizational misconceptions or distorted organizational images, sap the digital transformation efforts in public organizations, mainly by paralyzing the public workforce and the public organization. We find that work practices (e.g., restrictive thought patterns) contribute to gridlocks in public organizations, because rather than actively shaping the transformation, employees report to react to changes, for example by passively accepting technological advancement or by diffusing responsibility of decisions to other hierarchical levels. This pattern is witnessed not only at the individual level but also at the organizational level where misconceptions about digital transformation prevail. Changing these thought patterns requires a fundamental shift in world views from being a passive reactor to becoming an actor. This result is in line with research on digital transformation in general [48,49, e.g.,] that suggests that transformation includes changes in organizational identity and changed value propositions, but also requires organizations to actively use technologies to achieve such a change in value propositions. According to Wessel et al. [49], this process imposes changes in work practices. In a similar vein, Vial [48] suggests that structural changes need to take place for digital transformation to happen, among which are employee roles and skills but also organizational structures. In our case, we see that many of the process steps in digital transformation are impeded by obsolete competences that, therefore, need to be discarded systematically at both the organizational and the individual level. The insights generated in this study are therefore complementing and integrating results of similar studies, [11,31,39, e.g.,] by providing a comprehensive framework identifying which specific competences need to be unlearned. Hence, we underline the necessity of designing trainings to unlearn obsolete competences [12].

Aside from visible competences, we also find that many of the competences that public organizations need to unlearn are hidden, that is these are cognitive rather than behavioral competences [9]. Scholars repeatedly address the tensions that arise from cognitive competences that, on the one hand, guarantee continuity and, on the other hand, sap transformation. These competences are deeply engrained in the organizational culture and identity [14]; thus, they may be harder to unlearn. At the same time, we see that research on digital government competences mostly revolves around behavioral competences [13,38, e.g.,], and less around these hidden, cognitive competences. Similarly, educational programs designed for imparting digital government competences mainly target the behavioral ones [20,26, e.g.,]. Whereas imparting behavioral competences, such as process management, may increase the "employability" of individuals, competence development needs to address the "long-term building [of] relationships between employee and organization" [9] – a context-sensitive endeavor that can hardly be passed on solely in traditional educational settings.

Finally, our analysis is indicative of the complex and reciprocal relationship between individual employees' competences and the organizational core competences. Repeatedly, our respondents had difficulty in unambiguously assigning competences to either the individual or the organizational level – in their experience, both levels are closely intertwined. We find that, for example, public servants' self-concept may counter-act digital transformation efforts (individual level). Self-concept refers to an individual and how they see themselves in the organization; in our analysis it comprises *impeding attitudes* and *organizational misconceptions*. Yet, the roots of such aspects of the self-concept may also lie in the organization's culture that is built on values and work practices shared by the organization's members [14]. Analytical frameworks such as the one we chose to use in this work [9] suggest clear-cut borders between the individual and organizational levels as well as hidden and visible dimensions of competences. However, in reality, these borders become heavily blurred and hard to disentangle from the employees' perspective.

5.2 Concluding Remarks

While this research contributes to existing research on digital government competences, it also comes with limitations. For our research, we conducted focus groups with employees in German public administrations at the local level. The specific context of our research needs to be considered when transferring results to other contexts. Germany's public sector has unique characteristics that may have impacted the results. As the field of unlearning is still in an evolving phase [2], we had good reason to opt for an explorative and qualitative approach. However, we also acknowledge that our sample comes with some restrictions. We used five focus groups at the local administrative level that generated a large amount of material. Still, findings are restricted by a rather confined sample size. Future studies must determine how far insights from this study can be applied to other levels of administration and other contexts.

Thus, our analysis opens the path for further research on unlearning in public organizations. First, we see a clear need for more research on the interplay between individual and organizational competences. While the framework by [9] proved useful in guiding our analysis, its applicability is limited by the clear delineations between competence dimensions. The complex nature of public organizations may not be adequately captured, wherefore we suggest further in-depth investigations. More detailed studies can unearth the hidden competences affecting digital transformation and can explain competence transfer between individual and organizational levels.

Initially, we employed a deductive approach based on Chen and Chang's theoretical framework. Yet, engaging with municipal administration employees revealed challenges in aligning their feedback with our model. This discrepancy points to the value of more inductive methodologies, like ethnography or grounded theory, and design-oriented methods such as Design- or Action Research, to capture unlearning's dynamic nature more accurately.

Furthermore, our findings advocate for a reevaluation of competence management and learning in the public sector. The focus should shift from solely enhancing existing or acquiring new competences to supporting organizational change by facilitating the unlearning of obsolete competences. Embracing this shift is vital for public administrations to effectively navigate digital transformation, promising extensive societal benefits.

In this study, we aimed to pinpoint competences that hinder digital transformation in public administrations, identifying 24 competence categories in need of unlearning across organizational levels. Utilizing Chen and Chang's [9] theoretical framework, we adopted a deductive approach. Focus groups with public servants helped us empirically identify these competences. The qualitative analysis highlights the reciprocal relationship between individual and organizational competences and offers exciting opportunities for future research.

Acknowledgements. This research has received funding from the Deutsche Forschungsgemeinschaft (DFG, German Research Foundation) - FOR 5393, Project No. 462287308 (BE 1422/28-1; DI 2760/1-1).

References

1. Akgün, A.E., Byrne, J.C., Lynn, G.S., Keskin, H.: Organizational unlearning as changes in beliefs and routines in organizations. J. Organ. Change Manag. **20**, 794–812 (2007)
2. Becker, K.: Organizational unlearning: the challenges of a developing phenomenon. Learn. Organ. **26**, 534–541 (2019)
3. Bettis, R.A., Prahalad, C.K.: The dominant logic: retrospective and extension. Strateg. Manag. J. **16**, 5–14 (1995)
4. Bittner, E.: The concept of organization. Soc. Res. (1965)
5. Bloom, B.S., Engelhart, E.J., Furst, W.H., R., K.D.: Taxonomy of educational objectives: the classification of educational goals. In: Handbook I: The Cognitive Domain. Longman (1956)
6. Brinkmann, S.: Qualitative Interviewing. Oxford University Press, Oxford (2013)
7. Casey, A.J., Olivera, F.: Reflections on organizational memory and forgetting. J. Manag. Inquiry **20**, 305–310 (2011)
8. Cegarra-Navarro, J.G., Wensley, A.K., Sánchez Polo, M.T.: A conceptual framework for unlearning in a homecare setting. Knowl. Manag. Res. Pract. **12**, 375–386 (2014)
9. Chen, H.M., Chang, W.Y.: The essence of the competence concept: adopting an organization's sustained competitive advantage viewpoint. J. Manag. Organ. **16**, 677–699 (2010)
10. Cubillas-Para, C., Cegarra-Navarro, J.G., Wensley, A.: Unlearning as a future challenge for knowledge management. In: Bratianu, C., Handzic, M., Bolisani, E. (eds.) The Future of Knowledge Management. IAKM, vol. 12, pp. 149–168. Springer, Cham (2023). https://doi.org/10.1007/978-3-031-38696-1_8
11. Danneels, L., Viaene, S.: Simple rules strategy to transform government: an ADR approach. Gov. Inf. Q. **32**, 516–525 (2015)
12. Di Maria, M., Walter, D., Schoormann, T., Knackstedt, R.: Practical support for unlearning - a systematic review to organize the field. In: Proceedings of the 31st ECIS, Kristiansand, Norway (2023)
13. Distel, B., Ogonek, N., Becker, J.: eGovernment Competences revisited – a literature review on necessary competences in a digitalized public sector. In: Wirtschaftsinformatik Proceedings (2019)
14. Dull, M.: Leadership and organizational culture: sustaining dialogue between practitioners and scholars. Public Adm. Rev. **70**, 857–866 (2010)
15. Easterby-Smith, M., Lyles, M.A.: In praise of organizational forgetting. J. Manag. Inquiry **20**, 311–316 (2011)
16. Fiol, M., O'Connor, E.: Unlearning established organizational routines - Part I & II. Learn. Organ. **24**, 13–29 (2017)
17. Flak, L.S., Hofmann, S.: The impact of smart city initiatives on human rights. In: EGOV-CeDEM-ePart (2020)
18. Grisold, T., Klammer, A., Kragulj, F.: Two forms of organizational unlearning: insights from engaged scholarship research with change consultants. Manag. Learn. **51**, 598–619 (2020)

19. Grönlund, Å., Hatakka, M., Ask, A.: Inclusion in the E-service society – investigating administrative literacy requirements for using E-services. In: Wimmer, M.A., Scholl, J., Grönlund, Å. (eds.) EGOV 2007. LNCS, vol. 4656, pp. 216–227. Springer, Heidelberg (2007). https://doi.org/10.1007/978-3-540-74444-3_19
20. Halsbenning, S., Koddebusch, M., Niemann, M., Becker, J.: How to foster e-competence in the public sector? A mixed-method study using the case of BPM. In: EGOV-CeDEM-ePart (2021)
21. Haug, N., Dan, S., Mergel, I.: Digitally-induced change in the public sector: a systematic review and research agenda. Public Manag. Rev. **26**, 1963–1987 (2023)
22. Hunnius, S., Paulowitsch, B., Schuppan, T.: Does e-government education meet competency requirements? An analysis of the German university system from international perspective. In: HICSS Proceedings (2015)
23. Kitzinger, J.: Qualitative research: introducing focus groups. Br. Med. J. **311**, 299–302 (1995)
24. Klammer, A., Grisold, T., Nguyen, N., Hsu, S.: Organizational unlearning as a process: what we know, what we don't know, what we should know. Manag. Rev. Q. (2024)
25. Klammer, A., Gueldenberg, S.: Unlearning and forgetting in organizations: a systematic review of literature. J. Knowl. Manag. **23**, 860–888 (2019)
26. Koddebusch, M., Halsbenning, S., Kruse, P., Räckers, M., Becker, J.: The increasing e-competence gap: developments over the past five years in the German public sector. In: Fui-Hoon Nah, F., Siau, K. (eds.) HCII 2022. LNCS, vol. 13327, pp. 73–86. Springer, Cham (2022). https://doi.org/10.1007/978-3-031-05544-7_6
27. Lahti, R.K.: Identifying and integrating individual level and organizational level core competencies. J. Bus. Psychol. **14**, 59–75 (1999)
28. Lei, D., Slocum, J., Pitts, R.A.: Designing organizations for competitive advantage: the power of unlearning and learning. Organ. Dyn. **27**, 24–38 (1999)
29. Leonard-Barton, D.: Core capabilities and core rigidities: a paradox in managing new product development. Strateg. Manag. J. **13**, 111–125 (1992)
30. Martignoni, D., Keil, T.: It did not work? Unlearn and try again – unlearning success and failure beliefs in changing environments. Strateg. Manag. J. **42**, 1057–1082 (2021)
31. Matsuo, M.: Goal orientation, critical reflection, and unlearning: an individual-level study. Hum. Resour. Dev. Q. **29**, 49–66 (2018)
32. Mayring, P.: Qualitative content analysis: theoretical foundation, basic procedures and software solution. AUT (2014)
33. Mehrizi, M.H.R., Rodon Mòdol, J., Zafarnejad, M.: Managing obsolete knowledge: towards a clarified and contextualized conception of unlearning. In: ECIS Proceedings (2012)
34. Mehrizi, M.H.R., Romero Velasco, M.: A multiple psychological perspective of individual unlearning. In: Academy of Management Proceedings (2013)
35. Mergel, I., Edelmann, N., Haug, N.: Defining digital transformation: results from expert interviews. Gov. Inf. Q. **36**, 101385 (2019)
36. Mller, S.D., Skau, S.A.: Success factors influencing implementation of e-government at different stages of maturity: a literature review. Int. J. Electron. Gov. **7**, 136–170 (2015)
37. O. Nyumba, T., Wilson, K., Derrick, C.J., Mukherjee, N.: The use of focus group discussion methodology: insights from two decades of application in conservation. Methods Ecol. Evol. **9**, 20–32 (2018)

38. Ogonek, N., Distel, B., Becker, J.: Let's play... egovernment! A simulation game for competence development among public administration students. In: HICCS Proceedings (2019)
39. Pan, G., Pan, S.L., Newman, M., Flynn, D.: Escalation and de-escalation of commitment: a commitment transformation analysis of an e-government project. Inf. Syst. J. **16**, 3–21 (2006)
40. Peschl, M.F.: Unlearning towards an uncertain future: on the back end of future-driven unlearning. Learn. Organ. **26**, 454–469 (2019)
41. Polites, G.L., Karahanna, E.: Shackled to the status quo: the inhibiting effects of incumbent system habit, switching costs, and inertia on new system acceptance. MIS Q. (2012)
42. Prabowo, H.Y., Sriyana, J., Syamsudin, M.: Forgetting corruption: unlearning the knowledge of corruption in the Indonesian public sector. J. Financ. Crime **25**, 28–56 (2018)
43. Sinkula, J.M., Baker, W.E., Noordewier, T.: A framework for market-based organizational learning: linking values, knowledge, and behavior. J. Acad. Mark. Sci. **25**, 305–318 (1997)
44. Stahl, B.C., Tremblay, M.C., LeRouge, C.M.: Focus groups and critical social IS research: how the choice of method can promote emancipation of respondents and researchers. EJIS **20**, 378–394 (2011)
45. Tangi, L., Janssen, M., Benedetti, M., Noci, G.: Barriers and drivers of digital transformation in public organizations: results from a survey in the Netherlands. In: Viale Pereira, G., et al. (eds.) EGOV 2020. LNCS, vol. 12219, pp. 42–56. Springer, Cham (2020). https://doi.org/10.1007/978-3-030-57599-1_4
46. Tangi, L., Janssen, M., Benedetti, M., Noci, G.: Digital government transformation: a structural equation modelling analysis of driving and impeding factors. Int. J. Inf. Manag. **60**, 102356 (2021)
47. Tremblay, M.C., Hevner, A.R., Berndt, D.J.: Focus groups for artifact refinement and evaluation in design research. Commun. Assoc. Inf. Syst. **26**, 27 (2010)
48. Vial, G.: Understanding digital transformation: a review and a research agenda. J. Strateg. Inf. Syst. (2019)
49. Wessel, L., Baiyere, A., Ologeanu-Taddei, R., Cha, J., Blegind-Jensen, T.: Unpacking the difference between digital transformation and it-enabled organizational transformation. J. Assoc. Inf. Syst. **22**, 102–129 (2021)

Public Management Competencies in a Digital World: Lessons from a Global Frontrunner

Ulrik B. U. Roehl[1](✉) 📵 and Joep Crompvoets[2] 📵

[1] Copenhagen Business School, 2000 Frederiksberg, Denmark
ubur.digi@cbs.clk
[2] KU Leuven, 3000 Leuven, Belgium

Abstract. In contemporary public administration, mid-level managers play a key role in shaping if and how digital technology is used on every-day basis. But knowledge of the competencies of mid-level managers who increasingly need to manage use of digital technology is rather limited. Seeking to rectify the lack of knowledge, we distil relevant mid-level managerial competencies within thinking on bureaucracies, digital transformation and "digital skills", and combine this with a multiple case-study of Danish administrative bodies' use of automated decision-making. Drawing on an understanding of competences as a combination of work-related knowledge, skills and aptitudes held by individuals, we tentatively identify five key competencies for mid-level managers in relation to public administrative bodies' everyday use of digital technology. The identified competencies can help individual managers and public administrative bodies being more targeted in their efforts to manage increased technology use on an every-day basis. Noteworthy, none of the identified competencies imply managers should become technology specialists today or anytime in the future.

Keywords: Digital government · Mid-level management · Competencies

1 Introduction

More than 15 years ago, Dunleavy et al. [1] noted that the advent of the "digital era" was the most pervasive and structurally distinctive factor in changing governance arrangements and public administration in advanced industrial states. However, administrative bodies often fail to employ digital technologies in efficient and appropriate ways [2]. More recently, discussions of the impact of artificial intelligence (AI) on public administration has attracted considerable interest and authors have referred to both significant potentials and problems of technological, legal, political and ethical nature [3, 4].

With this development in mind, it is unfortunate that knowledge of the requirements that such increasing use of digital technology within public administration put on mid-level managers is rather limited.

Submission for the Digital Society Track for EGOV2024.

© IFIP International Federation for Information Processing 2024
Published by Springer Nature Switzerland AG 2024
M. R. Johannessen et al. (Eds.): ePart 2024, LNCS 14891, pp. 64–82, 2024.
https://doi.org/10.1007/978-3-031-70804-6_5

In hierarchical organisations, as public administrative bodies are at their core, mid-level managers play a key role. First, mid-level managers are directly responsible and accountable for the diverse activities of public servants "on the floor" (case workers, policy analysis staff, etc.) thereby exercising considerable influence over what is done and how it is done. Secondly, mid-level managers also embody the structural possibilities and constraints of public administration by navigating and balancing often conflicting demands in terms of bureaucratic standardisation and professional discretion [5].

Mid-level managers therefore play a key role in shaping if and how digital technology is used by public administrative bodies on every-day basis. In addition, mid-level managers also have an important role in negotiating and translating how expectations of increased technology use among superiors and external stakeholders are put into practice within their operational domain. At organisational level, mid-level managers inevitably influence the mentioned downsides and shortcomings of technology use as well as the potential upsides.

The lack of knowledge is all the more surprising as understandings within the social sciences of technology use are increasingly based on sociotechnical perspectives stressing that technology frames human possibilities for action but does not determine such action [6]. Digital technology within public administration does thus not exist independent of the public servants operating it or the organisational and cultural context surrounding it. Again, this assigns a key role to mid-level managers as they influence ordinary public servants' sensemaking and interpretations of technology use just as they themselves constitute an important element of this context.

Traces of relevant competences for mid-level managers in relation to every-day technology use within public administration can particularly be found in three streams of literature: 1) Classic Weberian understandings of bureaucracy [7]; 2) Recent contributions on how to successfully manage digital transformations within public administration [8]; and 3) Emerging contributions that discuss the necessary "digital skills" for public servants [9]. While those streams all represent valuable insights, none of them provide thorough, specific guidance to mid-level managers in relation to every-day use of digital technology. They either ignore digital technology; are rooted in understandings of technology induced organisational change as radical and rather infrequent; or focus overtly on the use of "hyped" technologies such as AI and blockchain.

While we build on the three mentioned streams of literature, our errand is to cast specific light on the necessary competencies of *mid-level* public administration managers to manage *every-day use* of digital technology. By focusing on every-day technology use, we wish to emphasise the daily use of increasingly complex ICT systems rather than sweeping transformations "from paper to bytes".

We specifically focus on mid-level managers pointing to their key role for the actual use of technology within public administration. Therefore, we delimit our analysis from requirements cast upon ordinary public servants, technology specialists and top-level managers (e.g., agency heads) within administrative bodies. We do not see requirements placed on those groups as inherently contradictory. But mid-level managers stand out in the sense that they, e.g., to a larger extent navigate the mentioned contradictory "double binds" [10] just as their platform for communicating compelling visions regarding technology use is often limited compared to top-level managers. Additionally, existing research on mid-level managers' broader roles within public administration tend to be limited [11]. In summary, mid-level managers play a unique role when it comes to technology use within public administration that beg a specific scholarly focus.

Drawing on Virtanen's [12] framework of public managers' competencies, we trace relevant insights from existing literature and combine this with rich, empirical data from a multiple case-study of use of automated, administrative decision-making within Danish administrative bodies in four different policy domains. Denmark has invested heavily in the digitalization of its public sector since the 1990s and is today considered a global frontrunner in terms of digital government [13].

Administrative decisions are central to public administration as it is through such legally binding decisions that public administrative bodies decide on what is lawful or not in specific cases in relation to individuals or firms. Increasingly, such decisions are automated by way of digital technology making use of automated, administrative decision-making a key example of daily use of increasingly complex ICT systems alongside, e.g., one-stop portals, open data initiatives and AI-based chatbots.

We focus on the requirements that increasing use of digital technology put on mid-level managers in public administration based on the following research question: *What are key competences for mid-level managers in relation to every-day technology use within public administration?*

Our focus is on the individual manager, and we understand competences as a combination of work-related knowledge, skills and aptitudes held by an individual [14]. Competences can roughly be perceived as an individual's personal learned capabilities as an alternative to more stable, personal virtues and character straits [15]. It follows that individual managers' competencies are susceptible to intentional as well as unintentional change, development and decay.

The contributions of the article are two-fold: Firstly, we tentatively identify five key competencies for mid-level managers in relation to every-day technology use that are interesting for both practitioners and scholars in the field. Theoretically, we pioneer a specific focus on the important role of mid-level managers in relation to technology use and stress that those do not need to become "tech nerds".

2 Existing Knowledge: Mid-level Managers and Digital Technology

We argued that mid-level managers are particularly important for technology use in public administrative bodies on an every-day basis. This might seem ironic, as scholars going back as far as Leavitt & Whistler [16] have consistently claimed that use of (digital) technology would ultimately make much mid-level management redundant. Such authors predicted that top-management will be able to significantly widen their span of control, as technology will provide information with greater speed, accuracy, and selectivity than mid-level managers can do. Following this, the numbers and role of mid-level managers would decline considerably. We, however, do not observe this trend in current West European public administration, just as earlier authors have repeatedly rejected it on empirical grounds [17].

2.1 Mid-level Managers: Characteristics and Competencies

Several authors observe that mid-level managers in public administration tend to be neglected in much research [18], just as the precise definition of the role is often unclear [19]. We understand mid-level managers as managers who exercise formal managerial authority (e.g., the deciding voice on hiring of new staff as well as of sanctioning existing staff). Contrary to some authors, we include "first-line" managers (managers directly managing ordinary staff), but exclude upper, senior and top-level managers with strategic responsibilities. Emphasising the importance of formal managerial authority, we also exclude staff with selective responsibilities for coordination, quality, finance etc. such as deputies, supervisors and team leads.

We further limit our analysis to mid-level managers within the "line". That is, managers who are directly concerned with organisational output-deliverables [19] whether those are making administrative decisions in the form of building permits, draft legislation or the provision of social services. Line functions and their managers are distinguishable from "staff" functions such as finance, human resources, and – not least – ICT, data, and other technological support functions. The activities line mid-level managers are responsible for are mainly organised in formal organisational units as opposed to the organisation-wide responsibilities of staff functions and upper and top-level managers[1].

Mid-level managers within the line share a number of characteristics that helps to explain their central role for technology use on every-day basis. They are, among others, accountable for the quality of their unit's output-deliverables, control expenditure and somewhat responsible for personnel well-being. Acting on behalf of top-level management, they exercise great control over the activities of ordinary staff but do so with procedurally-limited autonomy [19].

Reviewing articles on mid-level managers, Sudirman et al. [19] note that mid-level managers – obviously – have greater access to resources than ordinary staff but possess significantly less autonomy and control than top-level managers. Positioned at the crossroad between strategy and daily operations, they are key players in the translation of strategic plans, decisions, and intentions of top-level management and external decision-makers into operational activities. In this manner, they contribute to ongoing processes of strategic and organisational change even though they have little formal authority to act strategically. Besides the intentional transformation of strategy into operational activities, mid-level managers' sensemaking is important for projecting understandings and images of changed ways of working to ordinary staff [19].

As mentioned, scholarly work on public administrative mid-level managers' competencies has traditionally been limited. Writing on appropriate tools and techniques "to do an orderly and systematic job of managing", Drucker [20] noted that management research should not aim to provide managers (across industries and ranks) the right solutions but should focus on defining the right questions. Fast forwarding to digital transformation, Müller et al. [21] follow suit and note that digital transformation is a "multifaceted phenomenon with unique leadership requirements depending on *contexts*." [our italics].

---

[1] To avoid an overtly restrictive understanding of mid-level managers' responsibilities, we refer to their "domain of operations" thereby indicating that responsibilities of individual mid-level managers may extend beyond the organisational unit they head.

Reflecting the same tendency, one of the most cited frameworks on public managers' competencies stress the ability to turn relative general competencies into managerial action depending on internal and external contextual factors. Virtanen [12] thus introduces the five broad competence areas (categories) listed in Table 1: He argues that managers should ideally master all five areas, but the relative importance of the areas as well the nature of the exact competencies within each area depend on the context. Among several properties Virtanen tie to each area, "criteria of competence" is particularly helpful to illustrate the meaning of the competence areas. It is thus on the basis of those criteria that it can be assessed to what extent a manager masters a given competence area despite diverse contexts. Virtanen's framework is holistic and balanced, and provides a solid basis for analysing key competences for mid-level managers in relation to every-day technology use.

Table 1. Competence areas of public managers; adapted from [12].

Competence area	Description	Assessment criterion
Task competencies	Competencies making tangible "things happening" such as skills and behavioural techniques (e.g., data analysis)	Performance
Professional competencies in subject area	Specific competencies in the substantive field of the "line" organisation (e.g., social security) or in the organisation-wide "staff" function (e.g., ICT)	Development of the policy object
Professional competencies in administration area	General competencies in the execution of policy goals including the continuous improvement of execution	Development of policy execution
Political competencies	Competencies maintaining the formal and informal power to make real decisions that are politically acceptable	Legitimacy
Ethical competencies	Competencies in navigating and balancing contradictory goals and logics in ethically acceptable manners	Justification

2.2 Existing Understandings of Relevant Competencies

We trace existing understandings of relevant competences for mid-level managers in relation to every-day technology use within public administration in three streams of literature: Contributions rooted in classic, Weberian understandings of bureaucracy; recent contributions on how to manage and lead digital transformations within public administration; and an emerging body of literature that discusses the necessary "digital skills" within the public sector. Table 2 summarises the descriptions of relevant competencies and relates those to the competence areas of Virtanen [12].

Developing an understanding of the requirements that increasing use of digital technology put on mid-level managers, it is logical to start with the key characteristics of the public organisations that mid-level managers inhabit.

Today, we associate discussions on the special obligations and peculiarities of public administration with Max Weber's analysis of bureaucracy that position mid-level managers as superior officials overseeing the rational application of rules to individual cases [7]. It appears that most mid-level managers are still tied by extensive rules and procedures being at the receiving end of the scalar chain of authority from top-level management. While differentiating across public administration activities, the application of the general to the specific in a careful, transparent, and consistent manner thus continue to be a key element in much public administration that mid-level managers are primarily responsible for attending to.

A more recent perspective on mid-level managers and digital technology is focused on leading digital transformations within public administration. Digital transformations are here understood as the relative radical introductions of new ways of working closely intertwined with digital technology use and fostering new organisational identities [22].

The underlying assumption is the need for both high-level and mid-level managers to lead the change from one plateau of (less advanced) technology use to another plateau (of more advanced use). Here the ability to show active leadership including the ability to communicate a clear vision, ascribe meaning to changed ways of working and show persistence as setbacks build during technological change process are singled out as important managerial competencies [8, 23]

A third helpful perspective is the increasing number of contributions that – particularly within the disciplines of Public administration, Information systems and Human resources management – discuss the necessary digital skills within the public sector [24]. Often, the underlying sometimes implicit current is an ambition to further technology use within public administration and ensure its "efficient" use.

Across those contributions, two conclusions stand out: 1) Relevant competencies are not limited to the technical domain. Instead, the ability to combine understandings of digital technology with knowledge of organisational processes and political contexts are stressed [9]; 2) authors seem to agree that there is a lack of not only technological skills but also of combinations of the skills just mentioned [24]. Those contributions tend, however, to be tied to use of specific technologies such as AI in the form of machine learning [25] and seldomly discuss specific groups of public servants thus indirectly arguing that the requirements that increasing use of digital technology put on ordinary public servants and top-level managers are the same.

Table 2. Relevant competencies in existing streams of literature; main competence area refer to [12]

Literature stream	Key competencies	Main competence area(s)
Weberian bureaucracy	• Ability to ensure quality and legality of technology use • Ability to ensure consistency and balance in technology use • Ability to accept formal responsibility for technology use	Professional competencies in administration area
Digital Transformation	• Ability to communicate a clear vision for technology use • Ability to ascribe meaning to changed ways of working due to technology use • Ability to demonstrate persistence as setbacks and problems arise in relation to technology use	Professional competencies in administration area
Digital Skills	• Ability to operate digital technology at a basic level exhibiting a positive attitude to its use • Ability to conceive real-life in terms of data, probabilities etc • Ability to understand and promote potentials of greater technology use • Ability to integrate issues of digital technology with considerations of organisational, political, and ethical contexts	Task competencies; Political competencies; Ethical competencies

2.3 Empirical Requirements for Mid-level Managerial Competencies

Based on the theoretical understandings of relevant competences, this subsection introduces six organisational capabilities that illustrates the empirical requirements for mid-level management in relation to every-day technology use. The organisational capabilities have been identified in a recent study among Danish public administrative bodies on use of automated, administrative decision-making and good administration [26]. Simple said, a capability is an organisation's capacity to deploy resources to achieve a desired outcome [27]. The six capabilities thus describe administrative bodies' capacity to align use of automated, administrative decision-making with good administration. They do, however, also provide a basis for identifying requirements for mid-managerial competencies beyond the narrower focus on good administration.

Capability A, *Giving accurate and comprehensible reasons*, regards administrative bodies' ability to transform not only the logic of complex algorithms of automated decision-systems but also the detailed, high-volume data that often form the basis of automated, administrative decisions into comprehensible reasons for the "outside world" as embodied by the citizen or firm (the addressee) that the administrative decisions concern.

Capability B, *Informing addressees' expectations*, regards administrative bodies' ability to apply automated, administrative decision-making in a manner that both enhances the transparency of the decision process for addressees as well as providing them with a chance to correct possible faulty data held by the administrative body.

Capability C, *Combining material and algorithmic expertise*, regards how administrative bodies balance the need for "algorithmic expertise" (the ability to operate and understand automated decision-systems) with "material expertise" (the ability to navigate and understand the world of addressees).

Capability D, *Achieve effective oversight*, regards how mid-level managers are both helped and held accountable by their superiors to exercise effective oversight of automated decision-making.

Capability E, *Continuously ensure quality*, regards administrative bodies' ability to ensure effective procedures of continuous quality assurance of automated decision-processes.

Capability F, *Managing high complexity*, regards the ability of administrative bodies to manage the wider algorithmic systems (neighbouring ICT systems, databases, citizen portals etc.) in which automated decision-systems are typically embedded. Within those wider systems, both public and private partners almost constantly adjust and change "petite" elements that – worst case – can have a direct impact on automated decision-making of other administrative bodies.

The capabilities might look simple on the surface but are significant in the sense that they all cut across technology, working practices and organisational structures. It follows, that while they are unlikely to be perceived as the sole responsibility of mid-level managers in any organisation, mid-level managers are central to strengthening all six of them.

Assessing them in the perspective of relevant competencies for individual mid-level managers, the capabilities indicate the need for mid-level managers to draw on resources outside their immediate "line chain" from top-level management via themselves to individual staff. Relevant resources are not only financial (supplementing the ordinary budgets of managers) but also knowledge of, and capacity related to, technology use inside and outside the public administrative body. The identified capabilities thus refer to the importance of establishing fruitful cooperative relationships with managerial technology specialists internally (e.g., the chief information officer of the administrative body) as well as possible key individuals from relevant external providers of technology.

Another important realisation of the mix of technology, working practices and organisational structures for mid-managerial competencies, is the importance of being able to combine technological perspectives with traditional managerial perspectives on the necessary skills and qualifications among staff, adjustment of work practices and processes, redesign of organisational structures etc. Capability C, *Combining material and algorithmic expertise*, for example, implies an ability for mid-level managers to secure working practices that sufficiently and appropriately relate the algorithmic world (data, automated results, images etc.) with the material world (biological, physical, and social systems outside the administrative body).

3 Methods

The empirical basis of this article is based on a multiple case-study [28] of four policy areas in which Danish public administrative bodies employ semi and fully automated administrative decision-making: Illness benefits, Work retention, Agricultural subsidies, and Property value assessment. The design is a combination of what Yin [29] labels holistic and embedded designs. Distributed across policy areas and levels of government but representing comparable contexts of public administrative bodies highly reliant on technology use, the design provides a foundation for the analysis of relevant competencies. The design including descriptions of cases is illustrated in Table 3.

The case-study originally provided the basis for the identification of six organisational capabilities necessary to achieve supportive relations between use of automated, administrative decision-making and so-called good administration [30]. The identified capabilities are, however, also a rich empirical basis for identifying requirements for mid-managerial competencies beyond the narrower focus on good administration. Use of automated, administrative decision-making can thus be considered as an example of every-day technology use. In a similar vein, the ability to achieve supportive relations to good administration can be considered illustrative for mid-level management requirements in relation to every-day technology use.

Administrative decision-making and the resulting administrative decisions are central to public administration as it is through such legally binding decisions that public administrative bodies decide on what is lawful or not in specific cases in relation to individuals or firms. Increasingly, those decisions are based on digital technology ranging from semi-automated (the final decision being made by a public servant utilising assistance from automated decision-systems) to fully automated (the final decision being made by automated decision-systems) [31]. Automated decision-systems include, but are not limited to, techniques such as robotic process automation, rule-based (expert) models, regression, big data, predictive analytics, machine learning and neural networks.

In general, Denmark is considered a high trust society characterised by a digitalized, extensive and decentralised welfare state as well as social cohesion, low corruption and high equality [32]. Although differentiating in exact set-up, the administrative bodies of the cases are relying on products and services of commercial ICT suppliers.

Empirical data was collected in relation to each case in the form of interviews, documents, and observations as described in Table 4:

- *Interviews* were the primary form of data and were made with top and mid-level managers, specialists and caseworkers of the administrative bodies. All interviews were semi-structured, structured according to the position of the interviewee, took 45–90 min and were conducted physically or by video by the first author.
- *Documents* included internal guidelines and checklists for use of decision-systems as well as decision processes, public fact sheets, software documentation, internal teaching material, examples of decision-system templates etc.

- *Observations* were used as a supplementary data source via shadowing and stationary observations of caseworkers of administrative bodies [33].

A thematic analysis [34] based on within-case and cross-case analysis [35] was carried out searching for patterns in the empirical data. We applied Boyatzis' [34] hybrid approach, which blends inductive coding with existing theoretical assumptions in the identification of themes in the data. Interview transcripts, documents and observation notes were imported into "NVivo" (v. 12) qualitative data analysis software, where short summaries were created for relevant sources using the "memos" functionality.

Table 3. Case study design including description of cases; reproduced from [26].

Case	Focus of administrative decision-making	Circumstances of usage
1. Illness benefits • Municipality 1A • Municipality 1B	Determination of citizens' and firms' access to illness benefit reimbursements during periods of long-term illness (serve as basis for citizens' rights and duties of public benefits, compulsory job training etc.)	Approx. 15 public servants operate new decision-system. Decision-system is fully automated and public servants solely step in when individual cases are complex or essential data are missing
2. Work retention • Municipality 2	Determination of citizens' obligation to be available for employment after periods of long-term illness which influence citizens' rights and duties of public benefits, compulsory job training etc	Approx. 120 public servants operate well-established decision-system. Decision-system suggests relevant procedural steps for public servants
3. Agricultural subsidies • Government agency 3	Determination of farmer's access to agricultural subsidies	Approx. 50 public servants operate well-established decision-system. Decision-system is fully automated and public servants solely step in when individual cases are complex or essential data are missing
4. Property value assessment • Government agency 4A • Government agency 4B	Determination of the value of homeowners' real estate which serve as basis for the annual property tax	Approx. 500 public servants will operate new decision-system (in Government agency 4B). Decision-system is fully automated and public servants solely step in when individual cases are complex or essential data are missing

Table 4. Numerical overview of empirical data; reproduced from [26].

Case	Interviews	Documents	Observations
1. Illness benefits	21	21	7
2. Work retention	9	56	2
3. Agricultural subsidies	17	37	2
4. Property value assessment	13	45	2

Initially, this gave rise to the identification of six organisational capabilities necessary to achieve supportive relations between use of automated, administrative decision-making and good administration [26]. Together with the relevant competencies from existing literature, the identified capabilities also provided a meaningful empirical basis for identifying requirements for mid-managerial competencies beyond the narrower focus on use of automated, administrative decision-making and good administration.

The identification followed a traditional "Subsuming particulars into the general" principle [35] combining the specific competencies from existing literature with the requirements stemming from the specific organisational capabilities. Going back and forth, this eventually gave rise to the five key competencies for mid-level managers in relation to technology use.

4 Findings: Five Key Competencies for Mid-level Managers

Figure 1 combines the 10 competencies from existing literature with the requirements stemming from the organisational capabilities identified in the case-study. The result is five key competencies for mid-level managers in order to manage every-day use of digital technology (following Subsect. 2.1, the key competencies all have the operational domain of a given mid-level manager as their primary scope).

Competence 1 describes *the ability of the mid-level manager to envision possible applications of digital technology to support policy goals* within her/his operational domain. The competence regards the ability to conceive and communicate short-term and long-term use of technology. While conceptions and views of technology use is likely to originate from other sources, it is nevertheless the mid-level manager that is responsible for the use. More than being able to draw up specific ideas of technology use, the competence thus regards being receptive to ideas of applications of digital technology from others as well as to transform and compelling communicate those ideas to not only ordinary staff but also superiors. The competence is particularly related to thinking within the Digital Transformation/Skills literature streams, and is a higher-level illustration of organisational capabilities A and B from the case-study.

Competence 2, *continuously adjust skills, processes, and organisational structures to use of digital technology (and vice versa)*, is based on the basic observation that digital technology is and increasingly will be an absolute integrated part of public administration. Appropriate use of technology is therefore highly dependent on an effective alignment with staff skills, informal and formal work processes, and organisational structures.

Mid-level managers should therefore be capable of analysing and redesigning the relations between those. While this might sound abstract, it could be as banal as paying attention to the need for both on-job and off-job training of staff to maintain knowledge of appropriate use of key ICT systems. But it could also – with reference to organisational capability C from the case-study – be to have a long-term plan to ensure the correct balance of material and algorithmic expertise among staff including recruitment of new staff members and dismissal of existing staff. The competence is particularly related to Weber's thinking on bureaucracy, and concerns the importance of ensuring quality, legality and consistency in the activities of the bureaucracy as well as balancing the use of technology with other legitimate objectives.

Competence 3, *actively take responsibility for technology use*, to some extend builds upon competence B, and regards the ability and aptitude of the mid-level manager to actively take responsibility for use of technology within their operational domain upon themselves. Again, this builds on the basic observation that digital technology is and increasingly will be an absolute integrated part of public administration. Despite possible personal inclinations, lack of knowledge, disinterest, responsibility for the function of technology is anchored with the manager, and cannot be neglected or outsourced without comprising age-old norms of accountability. Again, this competence is particularly rooted in Weber's thinking on bureaucracy and is a higher-level illustration of organisational capabilities D and E from the case-study.

Competence 4, *the ability to position technology use within organisational, legal, political, and ethical contexts*, is more abstract and is likely to be harder to "pin down" in practice. The competence is rooted in the recognition that all public administration takes places within organisational, legal, political, and ethical contexts. The competence is particularly visible in the Digital Skills literature stream as authors increasingly stress how use of AI must respect both formal and informal ethical standards [3]. While not directly related to the organisational capabilities of the case-study, competence 4 is, nevertheless, illustrated by case 3 of the case-study that concerns use of automated, administrative decision-making by an administrative body in order to determine farmers' access to agricultural subsidies. Farmers have traditionally been a relatively powerful political group in Danish society who can yield considerable political influence when things are perceived as not going their way [36]. The implication for the mid-level managers in question is that they must somehow ensure that the overall design of technology use is not perceived overtly negative (in the sense of unwanted technical requirements) by large groups of farmers.

Competence 5, *the ability to utilise resources regarding use of digital technology within and across hierarchical and organisational structures*, is the most instrumental of the five key competencies as it can be seen as a basis for other competencies. It mirrors that management of technology use very often necessitates a blend of knowledge and insights that few individual mid-level managers fully command. Mid-level managers therefore need to be able to reach out to and utilise resources (knowledge, financing etc.) in order to support their own work. Such resources can be extraordinary funding for, e.g., training of staff or actual technology, but is just as likely to be access to knowledge and experience from sources and partners within and beyond the administrative body of the manager. Interestingly, this competence seems so fundamental that it is hard to

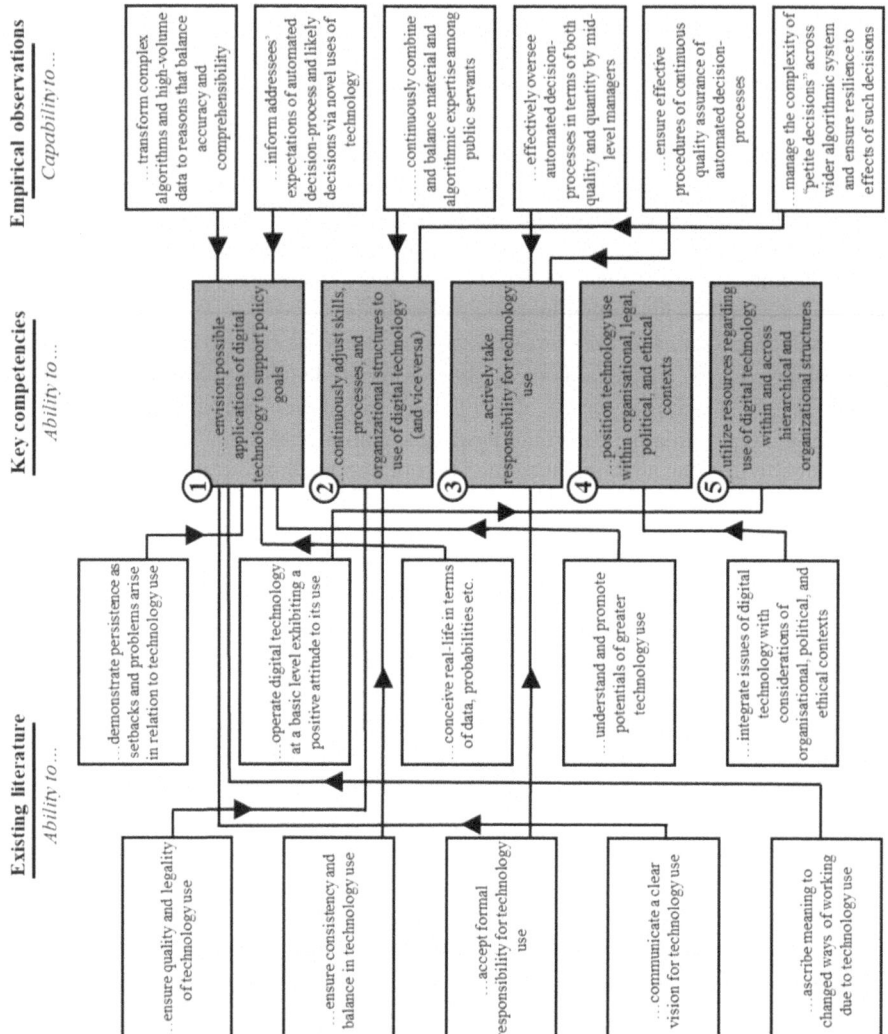

Fig. 1. Five key competencies for mid-level managers to manage every-day use of digital technology

find clearly described in existing literature. We do, however, see traces of the competence in the case-study, as several mid-level managers report examples of liaison with, and utilisation of, multiple partners both within and across existing hierarchical and organisational structures.

5 Discussion

This section discusses the common characteristics of the key competencies as well as their relations to existing scholarly work on roles and competencies of mid-level managers within public administration.

The above-mentioned key competencies do not emphasise mid-level managers' own use of digital technology. Seen across, more abstract aptitudes held by managers rather than tangible skills in relation to every-day technology use are thus dominating.

In Sect. 2.1, Virtanen's [12] model of five competency areas for public managers was introduced. Looking across the identified key competencies, Competencies 1, 2 and 5 primarily refer to the area of what Virtanen terms professional competence in administration. Such competences concern the ability to effectively execute any given policy rather than knowledge in specific policy areas. Competencies 3 and 4 supplement this and cut across Virtanen's areas of political competence and ethical competence. Whereas political competence concerns the legitimacy and authorisation of goals and means of public policy, ethical competence concerns the acceptability and justification of goals and means. Mirroring the lack of emphasis on mid-level managers' own use of digital technology, the identified key competencies do not include competencies within Virtanen's [12] task competence area that stresses the "use of instruments" and short-term performance.

Following this, the identified competencies do not necessitate that public managers display all required knowledge, skills, and aptitudes themselves. Instead, much of the necessary knowledge and skills can be made available to managers through deliberate delegation to, and communication with, own staff, partners within the administrative body or relevant external partners.

In this way, we do not find managerial responsibility for every-day use of digital technology to be much different than several other managerial responsibilities where the solving of tangible tasks is delegated, but the *responsibility* lies with the manager. An element that is reflected in Competence 3. Existing research do, however, indicate that mid-level managers may struggle with oversight of technology use [37] thereby, in effect, undermining their formal responsibility in a Weberian sense.

We have previously pointed out that existing literature only to a limited extent differentiates between competency needs of different categories of public servants. We argued that to further existing knowledge and provide meaningful guidance for practice, it is necessary to differentiate mid-level managers from ordinary public servants, technology specialists and top-level managers.

Relying on Virtanen's [12] areas, appropriate competencies for ordinary public servants as well as technology specialists in relation to every-day use of digital technology can be expected to be more rooted in the task competence and the professional competence in subject area categories. Nevertheless, the detailed competencies will differentiate for the two groups, as ordinary public servants will need to be able to combine operation of digital technology with knowledge of a given policy area and its associated policy goals and means. Technology specialists will, to a lesser extent, have to rely on knowledge of a given policy area including policy goals/means, as their subject area is not policy areas but the application and use of technology itself.

Top-level management will, on the other hand, ideally have competencies that to, a large extent, mirror the identified mid-level managerial ones but be able to deploy those at a more strategic and aggregate level. Competence 2 – that describes the ability to continuously adjust skills, processes, and organisational structures to use of digital technology (and vice versa) – illustrates this. Here, top-level managers have a larger organisational scope to consider than subordinate mid-level managers. But the necessary ability to analyse and redesign relations between appropriate use of technology, staff skills, work processes, and organisational structures is basically the same. Again, it is worth emphasising that the competencies we speak about here are related to every-day use of technology: There might be more substantial differences between relevant mid-level and top-level managerial competencies when it comes to digital transformation as indicated by, e.g., Edelmann et al.'s [8] emphasis of "leadership" as a central competency.

Another element that separates the competencies of mid-level managers from the competencies of top-level management is mid-level managers' need to navigate contradictory "double binds" [10]. A likely double-bind when it comes to management of every-day technology use is the contradiction between top-level management's abstract expectations of "more technology" while the same top-level management de facto maintains a "zero-error culture" through internal actions, sanctions etc. [38]. Mid-level managers must seek to combine the identified key competencies as well as broader managerial competencies in order to manage such binds, and thereby avoid sub-optimal and inconsistent managerial and organisational practices in relation to technology use.

It is also worth considering if there is anything uniquely public to the five identified competencies. After all, Max Weber did not specifically focus on public bureaucracy as we understand it today but on the principles of bureaucratic organisations. We have argued that appropriate managerial practices are almost always context specific. It follows that the competencies must be adapted to the manager's situation, e.g., the organisational context of the administrative body, the current level of technology use, goals of the relevant policy area etc. And such an adaption also includes a consideration of whether goals are profit-driven or public policy-driven. So while such an adoption of the competencies will clearly reflect whether a mid-level manager is, e.g., responsible for urban planning activities in local government or for advanced customer advising in a private sector insurance company, the key competencies themselves appear just as relevant.

There is one exception to the rule, however, and that is Competence 4 regarding the ability to position technology use within organisational, legal, political, and ethical contexts. Political and ethical considerations are of course not confined to public administration and the public sector, but we believe that the ability to interpret and understand the consequences of technology use in terms of politics and ethics is considerably more important for mid-level managers in public administration than for such managers in commercial organisations.

Summing up, understanding and reflecting upon technology use is increasingly a new "language" that mid-level managers ideally should speak. This language, for example, regards their ability to see potentials as well as pitfalls of technology use, ascribe and communicate meaning to possible changed ways of working due to technology, and position technology use within organisational, legal, political, and ethical contexts. It is appropriate to see this as the equivalent to how expectations to mid-level managers concerning, e.g., financial management, strategic thinking and innovation support, have crept up the agenda over time.

It should be noted that the identified competencies are not meant to be exhaustive for mid-level managers: The competencies are relevant in relation to every-day technology use and therefore supplement broader management competencies both in general [39] and within public administration [40]. As an example, the identified competencies do not include the rather mundane ability to instruct employees in solving tasks or employ tools based on digital technology. This is so, as the ability "to instruct" is a key competency in broader management competencies.

6 Conclusions

In public administration, mid-level managers play a key role in shaping if and how digital technology is used on every-day basis. Additionally, mid-level managers have an important role in negotiating and translating how expectations of increased technology use among superiors and external stakeholders are put into practice. Mid-level managers are thus well positioned to lessen the potential downsides and shortcomings of technology use within public administration as well as harnessing its potential upsides.

Knowledge of the requirements that increasing use of digital technology put on mid-level managers is, however, limited. Combining existing academic thinking with rich empirical insights from a case-study of Danish administrative bodies, we have tentatively identified five key competencies for mid-level managers in relation to every-day technology use within public administration: 1) managers must ideally be able to envision possible applications of digital technology; 2) managers need to understand how staff skills, work processes, organisational structure and use of technology increasingly relate, and continuously be attentive towards the need for realigning those elements; 3) managers need to actively take responsibility for use of technology within their operational domain upon themselves; 4) managers need to be able to position technology use within organisational, legal, political, and ethical contexts; and 5) managers need to be able to utilise resources in terms of knowledge, financing etc. both internally and externally.

The findings point to understandings of technology use as a new "language" that mid-level managers must be able to speak: This language does not regard mid-level managers' own use of digital technology, nor does it require mid-level managers to become "tech nerds". Rather, the language requires the ability to see potentials as well as pitfalls of technology use, ascribe and communicate meaning to possible changed ways of working, and continuously consider the consequences of technology use for skills, work practices, and organisational structures within mid-level managers' operational domain.

Much current academic thinking goes into the consequences of ongoing advances of AI. While the consequences for public administration are still emerging, we believe that the identified competencies are likely as relevant in relation to use of AI in public administration as in relation to use of other digital technologies.

Similarly, we believe that the relevance of the five key competencies go beyond the use of automated, administrative decision-making by public administrative bodies not only in Denmark but also in other national contexts. Most issues related to use of automated, administrative decision-making are akin to issues surrounding other applications of digital technology within public administration. Again, the identified key competencies must actively be adapted to the manager's situation, but there is no indication that this cannot be done across different applications of digital technologies or in other administrative traditions where the level of digital technology use is lower than in Denmark.

We do, however, suggest future research on how the key competencies are adapted in different managerial contexts. There is a need to cast light on the relevance of the competencies as well as how they are adapted across individual traits of managers (e.g., educational background and experience), existing levels of technology use, policy domains, and national, administrative traditions. Additionally, it will be beneficial with studies assessing to what extent the competencies "do the job" for mid-level managers navigating increasing use of digital technology, or – most likely – must be combined with broader management competencies. Doing this via qualitative case studies in different contexts seems promising, just as quantitative studies across a high number of individual managers and public administrative bodies will likely be rewarding.

In terms of the practical implications of our findings, the identified key competencies could help take the "drama out" of increased technology use for many mid-level managers. Although strengthening the competencies will potentially be demanding for some individual managers and public administrative bodies but do not require mid-level managers to become technology specialists today or anytime in the future.

Consequently, the findings do not indicate a need for a large-scale replacement of existing mid-level managers in favour of new "tech minded" managers. Rather, our findings point to a need for continuous retraining as use of digital technology increase in public administration.

Similarly, the competencies could help both individual managers and public administrative bodies being more targeted in their efforts to manage increased technology use on an every-day basis. Managers could utilise the competencies by considering which of them they master and which they do not. Further, they need to take into consideration what arrangements in terms of delegation to specialists, cooperation with internal and external partners, and management support infrastructures that could help them rectify possible shortcomings.

The findings also point to the importance of top-level management and human resource units in public administrative bodies implementing support arrangements that help mid-level managers strengthening the five key competencies as well as adapting those to the contexts of a given administrative body. Related, the five key competencies provide clear inspiration for a strengthening of technology aspects of public administration education at both graduate and executive levels.

As argued throughout this article, mid-level managers play a key role in shaping if and how digital technology is used by public administrative bodies on every-day basis. To ensure effective and balanced use of such technology, it is therefore central that top-level management actively engage in and support mid-level managers in a systematic manner rather than leaving daily management of technology use entirely to the individual manager. While the former approach will likely help administrative bodies harvest the benefits of increased technology use on an every-day basis, the latter approach risks not only overlooking the potentials of technology use but also risks creating unintended negative consequences.

References

1. Dunleavy, P., Margetts, H., Bastow, S., Tinkler, J.: New public management is dead – long live digital-era governance. J. Public Adm. Res. Theory **16**(3), 467–494 (2006)
2. Kempeneer, S., Heylen, F.: Virtual state, where are you? A literature review, framework and agenda for failed digital transformation. Big Data Soc. **10**(1), 1–13 (2023)
3. Henman, P.: Improving public services using artificial intelligence: possibilities, pitfalls, governance. Asia Pac. J. Public Adm. **42**(4), 209–221 (2020)
4. Young, M.M., Bullock, J.B., Lecy, J.D.: Artificial discretion as a tool of governance: a framework for understanding the impact of artificial intelligence on public administration. Perspect. Public Manage. Gov. **2**(4), 301–313 (2019)
5. Tummers, L., Vermeeren, B., Steijn, B., Bekkers, V.: Public professionals and policy implementation. Public Manag. Rev. **14**(8), 1041–1059 (2012)
6. Lips, M.: Digital Government: Managing Public Sector Reform in the Digital Era. Routledge, New York City (2020)
7. Weber, M., Kalberg, S.: Max Weber: Readings and Commentary on Modernity. Blackwell Publishing, Hoboken (2005)
8. Edelmann, N., Mergel, I., Lampoltshammer, T.: Competences that foster digital transformation of public administrations: an Austrian case study. Adm. Sci. **13**(2), 44 (2023)
9. van Noordt, C., Tangi, L.: The dynamics of AI capability and its influence on public value creation of AI within public administration. Gov. Inf. Q. **40**, 101860 (2023)
10. Røhnebæk, M.T., Breit, E.: Damned if you do and damned if you don't: a framework for examining double binds in public service organizations. Public Manage. Rev. **24**(7), 1001–1023 (2021)
11. Chen, C.A., Berman, E.M., Wang, C.Y.: Middle managers' upward roles in the public sector. Adm. Soc. **49**(5), 700–729 (2017)
12. Virtanen, T.: Changing competences of public managers: tensions in commitment. Int. J. Public Sector Manage. **13**(4), 333–341 (2000)
13. United Nations. E-Government Survey 2022. Department of Economic and Social Affairs, New York City (2022)
14. Nordhaug, O.: Human Capital in Organizations: Competence, Training, and Learning. Scandinavian University Press, Oslo (1993)
15. Macaulay, M., Lawton, A.: From virtue to competence: changing the principles of public service. Public Adm. Rev. **66**(5), 702–710 (2006)
16. Leavitt, H.J., Whisler, T.L.: Management in the 1980's. Harv. Bus. Rev. **36**(6), 41–48 (1958)
17. Dopson, S., Stewart, R.: What is happening to middle management? Br. J. Manag. **1**(1), 3–16 (1990)
18. Mcconville, T.: Devolved HRM responsibilities, middle-managers and role dissonance. Pers. Rev. **35**(6), 637–653 (2006)
19. Sudirman, I., Siswanto, J., Monang, J., Aisha, A.N.: Competencies for effective public middle managers. J. Manage. Dev. **38**(5), 421–439 (2019)
20. Drucker, P.F.: "Management science" and the manager. Manage. Sci. **1**(2), 115–126 (1955)
21. Müller, S.D., Konzag, H., Nielsen, J.A., Sandholt, H.B.: Digital transformation leadership competencies: a contingency approach. Int. J. Inf. Manage. **75**(September), 102734 (2024)

22. Wessel, L., Baiyere, A., Ologeanu-Taddei, R., Cha, J., Jensen, T.B.: Unpacking the difference between digital transformation and it-enabled organizational transformation. J. Assoc. Inf. Syst. **22**(1), 102–129 (2021)
23. Mergel, I., Edelmann, N., Haug, N.: Defining digital transformation: results from expert interviews. Gov. Inf. Q. **36**(4), 101385 (2019)
24. Ogonek, N., et al.: Towards efficient egovernment: identifying important competencies for eGovernment in European public administrations. electronic government and electronic participation. In: Joint Proceedings of Ongoing Research and Projects of IFIP WG 8.5 EGOV and Epart, vol. 23, pp. 155–162 (2016)
25. Wirtz, B.W., Weyerer, J.C., Geyer, C.: Artificial intelligence and the public sector — applications and challenges artificial intelligence and the public sector – applications and challenges. Int. J. Public Adm. **42**(7), 596–615 (2019)
26. Roehl, U., Crompvoets, J.: Inside algorithmic bureaucracy: Disentangling automated decision-making and good administration. Public Policy Adm. 09520767231197801 (2023)
27. Piening, E.P.: Dynamic capabilities in public organizations: a literature review and research agenda. Public Manage. Rev. **15**(2), 209–245 (2013)
28. Miles, M.B., Huberman, A.M.: Qualitative Data Analysis: an Expanded Sourcebook, 2nd edn. Sage, Thousand Oaks, CA (1994)
29. Yin, R.K.: Case Study Research, 4th edn. Sage, Thousand Oaks, CA (2009)
30. Ponce, J.: Good administration and administrative procedures. Indiana J. Global Legal Stud. **12**(2), 551–588 (2005)
31. Roehl, U.B.U.: Understanding automated decision-making in the public sector: a classification of automated, administrative decision-making. In: Juell-Skielse, G., Lindgren, I., Åkesson, M. (eds.) Service Automation in the Public Sector, pp. 35–63. Springer, Cham (2022). https://doi.org/10.1007/978-3-030-92644-1_3
32. Andersen, R.F.: Trust in Scandinavia: findings from moving borders between Denmark and Germany. Scand. Polit. Stud. **41**(1), 22–48 (2018)
33. Czarniawska, B. Fieldwork techniques for our times: Shadowing. In: Qualitative Methodologies in Organization Studies II, pp. 53–74. Palgrave Macmillan, London (2017)
34. Boyatzis, R.E.: Transforming qualitative Information: Thematic Analysis and Code Development. Sage, Thousand Oaks (1998)
35. Miles, M.B., Huberman, A.M., Saldaña, J.: Qualitative Data Analysis: a Methods Sourcebook. Sage, Thousand Oaks, CA (2014)
36. Daugbjerg, C.: Uniting to meet challenges: Danish farm interest groups in the 21st century. In: Surviving Global Change? Agricultural Interest Groups in Comparative Perspective, pp. 71–89. Ashgate, Farnham (2005)
37. Røhl, U.: Automated, Administrative Decision-making and Good Administration: Friends, Foes or Complete Strangers? Aalborg Universitetsforlag, Aalborg (2022)
38. Bentzen, T.Ø., Torfing, J.: COVID-19-induced governance transformation: how external shocks may spur cross-organizational collaboration and trust-based management. Public Adm. **101**(4), 1291–1308 (2023)
39. Mintzberg, H.: The Nature of Managerial Work. Harper and Row, New York City (1973)
40. Cox, R.W., III., Buck, S., Morgan, B.: Public Administration in Theory and Practice. Routledge, New York City (2019)

For the Fun of IT– In Search for Sensemaking of Digitalization Training Program for Leaders in Schools and Pre-schools in Sweden

Elin Wihlborg[1] and Maria Spante[2]

[1] Unit of Political Science, Department of Management and Engineering, Linköping University, Linköping, Sweden
elin.wihlborg@liu.se

[2] Unit of Inforamtion System, Department of Business, Economics and IT, University West, Trollhättan, Sweden
maria.spante@hv.se

Abstract. The local practice of digitalization in schools is a key component of digital inclusion and literacy policies in most states. However, despite all good ambitions of educational digitalization there are several challenges when it comes to implementation. This paper will elaborate on and contribute to the understanding of implementation of digitalization in governmental settings using an inductive approach with a specific focus on responsibility and sensemaking in mandatory versus voluntary implementation requirements.

This paper presents a case study with a mixed-method approach to examine how a Swedish local government, in charge of public education, initiated and implemented a program to enhance the competence of all school leaders in leading digitalization. This program was undertaken despite differences in digital governance requirements in schools and preschools at the time. The case study illustrates the importance of connecting sensemaking theory in implementation processes to enhance the awareness of meaningful realization of digital government in local practices. Additionally, this paper highlights the methodological implications of such theoretical approaches.

Keywords: digital government · schools · management · sensemaking and sense-breaking · case study

1 Introduction

There are several reasons to implement digital tools in schools, from the training of students and children to become digitally included in society and prospective labor market, to arguments for more efficient teaching [1–4]. Education is a common public responsibility in most states and public schools make up a central part of government expenditures, promoting digital citizenship and media and information literacy as well as critical resistance [5]. When introducing digitalization in schools, it must make sense both for the users and the professionals –, the teachers, school leaders and other staff – to be able to reach the desirable goals of the changes. The costs are high, and it is complicated to evaluate if and how investments and training reach the targets, as shown in a recent review article on digital tools for children with special needs [6].

Despite identified benefits, organizing digital transformation for learning in school has proven to be a daunting and multi-layered task over the years. Previous studies argue that it is important to address issues such as culture and structure, as the potential of digital technology to become supportive of learning is closely linked to the context in which the school operates [7–9]. There are high policy ambitions both in the field of digitalization and education in most states, and combined they illustrate the main challenges of finding capabilities and functionalities of digitalization [10].

These policy ambitions for digitalization in schools is closely related to the discourse on digital inclusion, since digital inclusion has shown to be a key for social inclusion and justice [11]. Policy awareness of digital inclusion and digital literacy builds on research on digital divides [12]. The use of digital tools in schools addresses a complex and sometimes incoherent array of policy ambitions and relates to the broad policy ambition of increased digital inclusion that street level bureaucrats, as school leaders and teachers, have to address [13, 14].

However, when more general policies on digital inclusion and literacy are translated into educational practices there is a need to make sense of the policy idea in the educational context and the professional roles and competences of school leaders and teachers.

This paper, using a bottom-up approach, presents research results building on a process evaluation of a training program for improved use of digital tools in schools and pre-schools in a Swedish municipality. We use the case study as a starting point to elaborate on how participants make sense of the training program in their daily practices and thereby how professional competencies are integrated and combined into the educational curriculum to enhance students' and children's digital literacy and inclusion.

1.1 Aim and Research Questions

This paper aims to narrate how a leadership and competence development program changed the meanings of digitalization in schools. The analysis reveals how school leaders made sense of the implementation, and we elaborate on interpretations on sense-breaking to explain the different outcomes of the project.

To maintain our inductive reasoning, we first present the case results to elaborate on the methodological and theoretical implications. The paper is structured around three research questions.

– How did the school leaders make sense of the training program for digitalization in schools to improve their educational practices? (Sect. 3)
– How did our methodological approach envisage the sense-making of implementation? (Sect. 4)
– What theoretical lessons can be learned from the narratives told in the paper, and how can we elaborate on sense-making of digitalization? (Sect. 5)

The outline above indicates that this paper deviates from a typical structure. This is an attempt to inductively present an argument on how to look beyond conventional frameworks [15] to find theories and models to interpretate the effects of digitalization in governmental settings. This paper proceeds firstly with a brief research overview and contextualization of the case before addressing the research questions above. In Sect. 3 we empirically discuss the implications from the process evaluation as a case study. In Sect. 4 we reflect upon the methodological lessons learned. In Sect. 5 we elaborate on the theoretical implications. Finally, we draw some more general conclusions and suggest ideas for further research. This structure requires an open approach to see how the case study guides us towards the theoretical implications.

2 Research Overview and Background of the Case Context

In this section, a brief literature review on digitalization in schools is presented and we provide an overview on how schools in Sweden are managed and the specific discretion for municipalities to enforce participation in a training program as in this case.

2.1 Brief Research Review on Digitalization and Leadership in Schools

Digitalization in schools takes many forms, for example teaching through films and games, but also honing information seeking skills and creativity. Thus, digital tools can be seen as beneficial for the learning process, classroom interaction and, of course, the development of digital competences among pupils [2]. Not unlike the arguments for digitalisation of public service in general, digital tools in schools are also often described as a means to reduce the workload of teachers. Today, many schools administer their education through digital learning platforms [16], and acceptance to use digital tools is often measured by factors such as attitudes and experience [17].

Research underscores the importance to continuously evaluate how digitalization co-constructs education in all aspects from management to social activities, not at least the influence of the tech industries and its impact on equitable participation in education [18]. Grönlund [19] points out that effective digitalization in schools depends on effective implementation. Selwyn [18] further asserts that these significant concerns in digital government must be considered when making deliberate choices and decisions regarding the digitalization of educational systems to reduce risk of unintended consequences.

Previous research has shown that supportive leadership, supportive organizational culture and collegiality are singled out as important foundations for the professional development of teachers' digital competence [20]. In addition to teachers' competences to develop methods, supportive measures, and collaborative routines for students, school leaders need skills to support teachers in their professional development regarding school digitalization [21]. Vanderlinde and van Braak [22] stress that it is the school leader's role to create changes through the process of implementing digital technology in schools that are driven by top-down policy decisions. It is clearly the case that surrounding factors in combination with leaders' capacity to handle the implementations become important for success.

Despite the key role of school leaders, they receive little education in how they can support inclusive digitalization through their leadership [23]. Awareness of external pressure is crucial for school leaders since such pressure has shown to increase the risk of creating an overly instrumental and formal approach in leadership as a response to external pressure to meet measurable goals. Moreover, one paramount risk with overly instrumental and formal approaches is to harm meaningful learning environments in the school [24]. Furthermore, digital transformation has in research been identified as a common organizational challenge that school leaders often struggle to manage [25, 26].

2.2 The Context of the School System in Sweden

The case study presented in this paper is set in Sweden, where schools are governed by a national curriculum policy but managed locally by the municipalities (https://sweden.se/life/society/the-swedish-school-system). Preschools are optional and designed for children aged 1–5 years old and regulated by a national curriculum for preschools. Elementary school, known as primary school in Sweden, is publicly funded and compulsory until grade 9, age 6–16. The purpose of elementary school, according to the national curriculum, is to ensure that students acquire and develop civic competences and knowledges necessary for each individual and prepare them for eligibility for upper secondary school. Among other objectives, digitalization is addressed both as a media and a communications tool, and as a new context in which students have to develop competences to evaluate and manage information and influences regarding individuals and the society. The national curriculum further states that exploration, curiosity, and a desire to learn shall form the basis of the school's activities. The school shall offer students structured teaching under the guidance of teachers, both in whole-class settings and individually. Teachers shall strive to balance and integrate different forms of knowledge in their teaching.

Full public funding is provided for both pre-schools and schools, as well as for independent schools, and pre-schools, that are accredited by national agencies and funded by the municipalities on a voucher-like basis. This governance model has been extensively discussed for the quality of outcomes [27].

There have been several ambitious policy programs targeting digitalization in schools, such as the one-to-one program (one student–one computer) [28]. However, recently both research and policy makers argue that there are several challenges with the digital competences and literacy in schools in Sweden in general even if there is an increasing disparity between students who succeed in school and those who struggle [29].

3 Field Study Results

The field study was designed as a process evaluation following a training program for school leaders. This section presents the main results from the case study, which opens up for methodological interpretations in Sect. 4, and theoretical implications in Sect. 5.

3.1 Case Study Background and Field Work Data

The case study was initiated when one of the authors were invited to follow a training program on digital education for school leaders in a medium sized Swedish municipality. The program aimed to train school leaders to lead digitalization, building on the school leaders' skills and understanding of the multifaced and complex use of pedagogical and administrative digital tools. The program was compulsory for all school leaders in schools and pre-schools in the municipality. The curricula for the two levels of education are partly different, but with similar overall ambitions. Digitalization was mandatory in schools but voluntary in pre-schools.

The background to the local training program originated from formal decision by the local government council driven by a policy initiative. The decision was to make it compulsory for all school leaders in the municipality to take part in the national program "Leading digitalization" formed by the Swedish National Agency for Education. The main objective was to support leaders in schools to promote and use digital tools as much as possible and to promote digital literacy and collaborative skills among students and children.

The project was implemented through a series of training workshops, led by a facilitator, over one academic year, including all school leaders in the municipality. This took place the year before the covid-19 pandemic and thus no references are made to distance education, or related issues based on the pandemic constraints.

As researchers we got an opportunity to follow this program's implementation closely, by observing meetings, conducting surveys and interviewing key actors during the implementation phase, as well as conducting a follow up survey one year after the eighth month program period. The program was managed by the quality management unit at the central municipal administration. In total about 40 school leaders participated. In line with the general municipal organization and different curricula the school leaders in schools and preschools had separately arranged workshops. In each group smaller learning groups were formed for leadership training, and here we made participatory observations during the meetings where the participants reflected upon the change management process and their roles as leaders.

Two of these groups were followed more closely. In these two groups, one with school leaders and one with pre-school leaders, we observed how the facilitator guided them through the educational material and mentored in-between the monthly workshops. The workshops addressed challenging questions that the participants brought in to trigger organizational-related discussions and reflections.

In total we participated in and collected data from six sessions, each lasting 120 min, followed by a reflection meeting of 40 min with session leaders after each session. All participants gave an informed consent, for the process evaluation and related research on the topic. All quotations from are translated by the authors.

3.2 Different Meanings of Digitalization in Schools

During the field work it soon became clear that the participating school leaders had different views on what was seen as possibilities and constraints for a successful implementation of the digitalization program at their schools. They used different metaphors when talking about their school and how they translated lessons learned in the program into daily practices. The school leaders tended to talk about problems with specific systems and the lack of support from the municipality. The preschool leaders, on the other hand, tended to talk about how newbies learn, and suggested methods and approaches that involve children to develop an exploratory spirit among staff.

These reflections on the meanings and interpretations, as seen during workshops and in interviews, came in a context they described as open and trustful as one of the participants said:

We trust and support each other and our fantastic facilitator made us think one beyond the program

The differences that emerged during the observed workshops showed that the school leaders and the pre-school leaders framed their use of digital technology differently. They also articulated their experiences in different terms and included in a quite scattered palette of examples of digitalization – stretching from the learning platform to text-messaging with parents and the budget and staff planning programs. Their different views on digitalization were evident and we could observe a correlation to their leadership approaches. One group interpreted and talked about digitalization as integrated into pedagogical practices and organizational development. In the evaluation directly after the program one of them wrote:

This would also be a leadership training [opportunity] and sometimes the tasks have been too focused on digitalization and not integrated into our daily work.

These contrasting frames of digitalization for school leaders guided us into the focus on sense-making. We could see a need to uncover the making of meanings of digitalization as a dialogic process, where metaphors formed different frames of the expected changes as discussed, by relating to a similar interpretation of an analysis of sense-making of global societal transformations towards sustainability [30]. In the next section we extend the interpretations of sense-making.

3.3 A Gap in the Chain of Command

The other main impact of the process evaluation was the gap in the chain of command. The school leaders participating in the program had several challenges to motivate both participation and to implement what they learned in their daily practices. They expressed constraints to embed the new knowledge into their daily work. Already during the training program, the management team, at the central municipal office, identified these challenges and looked for ways to support the participants and adjust the program but with limited results since they were outside the lines of management as a project team.

The school leaders mainly highlighted the conflicting attitudes between their competences and the resources available to be accountable for this program in addition to their regular responsibilities. They expressed challenges in implementing and integrating these changes into their daily work and among teachers who are already burdened with their daily tasks. There was also criticism on the pace of the program, as one of the participants expressed:

I think things have been going a little fast, and I don't feel that.
I have been able to anchor all the information...
... it has been challenging to integrate and work with activities [in the school] as I wanted.

On the other hand, the preschool leaders were mainly positive to the program. The digital program was not mandatory for them, so there were more openings for local adoption and adjustments when integrating what they learned in their daily change management and leadership. The overall impression was that they came back to the saying – we do this for the fun of it. The integration of digitalization in the pre-school education gave an extra positive meaning, and they experienced support through their organization, as in this response:

There were several positive effects even upwards [in the chain of command].
when we as a group focused on this. For example, we got.
our digital tools paid in addition to our ordinary budget, we could take time for learning and reflection.

In line with this illustration, most leaders in pre-schools generally said they had freedom and competences to decide how and when to implement what they learned in the program and make changes for teachers and children in their schools.

These impressions point at the importance of knowing the context to make sense of the implemented changes. It indicates that there are more general impacts of the other often very contextual interpretation of sense-making.

3.4 Lessons Learned During the Program

In connection to the last workshop an evaluation questionnaire was distributed. In the answers to the free text questions, we could identify four themes. Firstly, digitalization was interpreted as something extensive, more important than previously thought of among the leaders in schools and pre-schools. Secondly, all of them had had challenging experiences and met challenges when bringing the new digital teaching approaches into daily practices. Thirdly, a common remark was that they were surprised of how rewarding and comfortable it was to share experiences in the group and get support from the facilitator. Fourthly, we could see a common thematic discussion on the visibility of digitalization and the competences that can be enhanced in a group of teachers, as one of the participants wrote in the evaluation: *"I'm surprised that the digital competence of the educators is so widespread."*

In the evaluation form, they were also asked to see to more long-term effects of the program. Here we could see that there was a strong desire to maintain the learning groups that have been supporting the daily work. The participants recognized the value of these groups in their ongoing work and wished for their continuation. Secondly, they wished to retain the facilitator, since the support they gained for daily challenges as school leaders working on digital transformation still were essential. They also highlighted their own needs to keep up learning new digital tools and platforms, to be able to gain competence to lead others. The evaluation form pointed at the need for school leaders to be digitally competent and confident to be able to lead teachers in this digital transformation.

3.5 Experiences After One year

To grasp more long-term effects of the program we conducted a follow up survey focusing on learning one year after the program. The questionnaire consisted of 30 questions relating to lasting effects and how they used the new knowledge in their daily practice at that time.

The most significant and overall impression was that the pre-school leaders still had positive impressions of the program and most of them still met to collaborate and learn from each other even without the facilitator. None of the school leaders reported that they kept meeting in their group. The pre-school leaders reported that they had continued to work on the lessons learned through the program and still strived to develop and implement more digital activities and learning tasks together with teachers and children.

In the follow-up survey we could see a form of social learning among school leaders was highly valued, and the pre-school teachers were most eager to keep this, for them new, networking and peer-support opportunities. The school leaders that had to follow the program and implement changes among their teachers and students were less positive. They requested more support and sustainable structures to be able to integrate what they had learned in the program. The training did not match their daily management. In the open questions they raised demands like:

– *improved capacity in reporting and talking about creative and efficient efforts could help transform projects,*
– *possibilities to reflect upon digital transformation to continuously create meaningful learning*
– *improved working conditions and competence development for all professionals in schools.*

3.6 Concluding Remarks from the Case Study

In line with our inductive approach, we now identify the empirical impacts that we will cultivate further in the coming sections.

The main impact is that school leaders in schools had to make use of the training and change the daily practices in their schools, and at the same time deal with an already heavy workload. Digitalization was already a key component in their work. The pre-school leaders, on the other hand, had more discretion to use and implement their new knowledge freely and in relation to their leadership approach. During the program, both groups of leaders shared experiences of relevance, possibilities, and enactment. However, the follow up study after one year showed divergence between the groups where school leaders expressed contrasting challenges that we saw as a form of sense-breaking with the program compared to pre-schools leaders that continuously expressed sense-making experiences in their practice at their pre-schools.

An important aspect, however, was how to proceed with efforts due to differences in attitude as it was generally expressed between school leaders in school compared to leaders in preschool. Among the school leaders, the program was seen as something with a beginning and an end, while among the preschool leaders the process continued but with a certain change of direction and a greater focus on continuous development and leadership.

The difference between the school leaders and the preschool leaders thus becomes important to pay attention to and to act upon. There is every reason for managers of the municipality to identify which types of differences are acceptable and which ones really need to be addressed. The constant reflection and discussion regarding digitalization is also seen here as central to the school as an organization [18, 19]. Of course, it is important to relate to the different school forms based on the assignments they have and at the same time keep the development of the different school forms in such a way that one does not risk associating digitalization with a technology project but insisting that it is about learning and student development where digital technology sometimes has an important part in learning.

4 Methodological Lessons Learned

This research's contribution builds on a re-analysis of a case study [31] opening for more conceptual and theoretical implications. Thus, it reflects upon the case study in an interactive evaluation approach, and some methodological lessons learned.

4.1 On-Going Evaluation and Collaborative Research Approaches

The strength of on-going evaluation research is the combined role of the researcher as an external observer who also can act as a critical friend during the process, as was the case in this field study. In this case we had several recurring occasions, in addition to the field study. On these occasions the researchers gave feedback to the management group and the policy makers on the school board. The aim of these feedback sessions was to be transparent with what was going on and what was observed and interpreted during the process, and to enhance the implementation process.

We have followed a collaborative research approach by formulating questions, activities and select methods jointly among practitioners and researchers. Nilsson and Sorbring [32, p. 19] describe collaboration research as "it is about research in collaboration between researchers and practitioners with the dual purpose of extracting both action base and theoretical knowledge and to develop and improve the current practice". This type of research has emerged as a response to societal challenges that require more collaboration to gain knowledge together regarding both improved practice and theoretical depth rather than to settle for distanced theorizing [33, 34].

In this study, we focused on the strength of the combination of collaborative research and on-going evaluation research with the aim of jointly asking relevant questions and getting closer to the effects that a competence development effort can generate. However, it is important to bear in mind that effects here are not about limiting themselves to studying key performance indicators when making such measurements on a continuous basis. In this study, it is intended to see how the practice changes, takes a new direction, and relates to statutory goals for school to enhance change management and professional training. Hereby, the theoretical implications discussed in the next section are clearly grounded in real experiences among the participants in the program. Our proximity to the participant increased the reliability of that discussion and thus we will discuss proximity as a methodological lesson learned.

4.2 Methodological Proximity

A difficulty with on-going evaluation research is the importance of dealing with both proximity and distance to the phenomenon being studied. There is a risk of becoming too uncritical of what is being studied by creating close relationships with those involved in the process during the study [36]. At the same time, these relationships also open opportunities for getting closer to the phenomenon.

The proximity enables collaboration and opens for a common understanding of what is happening and opportunities and obstacles that the participants meet when trying to achieve desirable goals. In addition, proximity helps to visualize difficulties and the diversity of opinions and presumptions that they must handle in their local context every day. By being close to the process differences in the researchers and participants experiences and understanding can be addressed throughout the research process. Instead, the differences can be seen as opening for learning and for continued progress. This in turn can, according to Albinsson and Arnesson [35] provide insights for participants on how to manage local conditions to reach the expected goals, and for researchers for further analyses.

In this case, our proximity was clearly achieved by being present at each group's discussions. The continuous presence opened for continuous interpretation of expressed experiences among the school leaders, and we learned to see beyond the specific context when we almost go to know the participants. Hereby, we could see what manifested as sense-making expressions in the collective discussions. Similarly, continuous proximity deepens conversations about the view of digitalization reducing risks and getting into limited conversations regarding the number of gadgets and systems, but rather to consider digitalization as organization processes that hopefully are driven in the direction of organization development and desirable transformation to create meaningful teaching and learning.

Given our methodological approach we could capture changes given the study design involving continuous participation and a follow-up. Moreover, we could also manage the need for distance with the study design, avoiding the risk of being too un-critical as a critical friend during the study [35].

Distance, from the studied program, was catered by capturing individual experiences in a survey with open-ended questions on the participants' experiences. However, an important note is that the dynamic combination between dimensions of proximity and distance was possible due to the on-going interaction with participants and actions. A static research design or survey only, would not have been able to capture all the nuanced experienced address during the process.

5 In Search of Theoretical Implications

In this inductive presentation of our case study, comparing outcomes in a program to enhance lead digitalization in schools and pre-schools, we have noticed two different outcomes of the same program. It shows that context matters [14] and thus we will elaborate on how theories of sense-making and show how context matters. This analysis elaborates on and discusses the concept of sense-breaking since we have noticed that the school leaders did not improve their leadership, rather lost what had earlier made sense to them as leaders of digital transformation in schools.

5.1 Sense-Making as a Theoretical Frame

Sense-making refers to the cognitive process through which individuals and groups create meanings and make sense of (complex) situations. The sensemaking processes involve interpreting information, constructing narratives, and developing shared understandings. Sense-making is particularly relevant when dealing with complex social issues where multiple perspectives, uncertainties, and ambiguities exist [36–38].

Sense-making has been used in analyses of similar cases of competence development and professional learning [39–41]. The concept of sense-making highlights the importance of a process approach rather than a focus on the result. This is especially relevant in digital transformation and as here when integrating multileveled organizational change management aspects, as seen in studies of technology acceptance measurements [17].

Sensemaking takes place in each context, thus an analysis of sensemaking has to understand the context. As a cognitive processes, sensemaking takes place in a specific social, cultural, and organizational context. Digital government research and practice has also been criticized for regularly over-emphasizing the importance of individuals and the unique institutional contexts, where transformation takes place [42].

The sense-making frame opens for diverse interpretations, values, and beliefs held by different individuals. Sense-making focuses on how challenges are being interpretated, if they make sense in the specific context. It can visualize the gaps between the intended policy or intervention and its actual enactment, and how different actors individually perceive and respond to these challenges and take account to others in the same context [43]. Thus, relational feedback loops are essential for sensemaking, since sensemaking needs a framework for understanding the mechanisms through which feedback loops occur. Feedback loops combine the implementation outcomes with the aims, and can shape subsequent actions and decisions, leading to adaptations or modifications in the implementation process. Sensemaking can inform adaptive strategies in a change process, as actors make sense of their experiences, they may identify areas that require adjustment, innovation, or learning. Implementation theories can help structure these adaptive strategies by providing frameworks and evidence-based practices [37]. Sensemaking builds on learning and knowledge sharing, when the processes proceed to make sense.

In this case learning and knowledge generation became very explicit, since we studied the learning for leaders during the process. By understanding how actors make sense of the program there is a potential to uncover insights during the process that could improve their leadership.

5.2 Does Digital Transformation Make Sense for the Leaders in the Program?

Although digital transformation is seen as a process, this program showed that it was important to know what you have, what you do and how it goes for each participant. All leaders had different backgrounds, competences and assignments. In the workshops they repeatedly came back to what made sense for them and why.

The training program, as well as the digital transformation they were supposed to lead in their schools, made sense in very different ways. Context did clearly matter. The professional conversations in the workshops showed that the training could get a positive feedback loop in the preschools. Here, leaders came back to each workshop with new insights and adapted strategies they had learned to their context. They found the fun of IT, to share with teachers and children and shared it in the workshops – as a feedback loop. On the other hand, in the schools where digitalization was not new but compulsory, the school leaders could not make sense of the training as leadership nor as a change management tool for digital transformation. The training in the workshops could not be implemented in their daily context and they could not interpret the challenges they met. The schools already had an extensive curriculum on digitalization and most teachers had their own practices based on different competences in relation to what and how they taught. The discretion for the school leaders was limited, in contrast to the openings in the pre-schools without a present curriculum om digitalization.

5.3 Sense-Breaking When There is no Fun in IT

In contrast to the pre-school leaders who brought in digitalization for the fun of it, the school leaders could not find meaning in the training program. For the school leaders the program was compulsory, and they already had extensive digitalization in the schools. For preschools leaders the program was voluntary, and digitalization was still new. Sensemaking activities, such as reflections and sense-giving interpretations of challenges influence how actors interpret and respond was clearly present among the preschool leaders, but not among the school leaders. In the case of school leaders, the program can rather be seen as a form of 'sense-breaking'. After one year they could not see any lasting consequences, responses or lessons learned. This was a form of sense-breaking.

Sense-breaking has been argued to be undertheorized, but Ybema and Willems [44] explored its disruptive and deconstructing strategies. They build their theory on an analysis of how public policies can break and even break-up common meanings of public organizations. Thus, sense-making and sense-breaking can be keys to analyze different outcomes and make contextual factors more visible. Even if there is a conceptual ambiguity in the very contextual term "sensemaking", by elaborating the on how a training program can be sense breaking if not contextualized enough.

The lack of a unified definition of sense-breaking can make it challenging to compare results and draw cohesive conclusions from the research. However, the concept of sense-breaking may relate to the fact that sensemaking always is seen in retrospective with a given outcome. The sense-breaking in schools may also come from a neglect of power relations, as in the line of command following a public policy steering. In schools, with teachers in stronger professions as carrier of their own curriculum and in addition engaged parents, there are more power relations to consider when trying to make sense of a new training and leadership. This in contrast to the pre-schools with a more open curriculum and flexibility in daily activities.

6 Concluding Discussion

This case study initiating a novel combination of how sensemaking can enhance and discuss implementation studies, includes several common challenges of implementation. It is uncommon that policy implementations are voluntary so what are then the implications of this observation from the comparative study? Our main conclusion is to draw upon the process view provided by the pre-school leaders linked to digital governance. We conclude that this approach would be beneficial also for mandatory implementations. Since sense-making has shown to be tightly coupled with the orchestration of implementations [39–41], there is a developmental gain to be fulfilled with a process approach rather than getting caught up with a project view that seems to become particularly present in mandatory model implementations. Therefore, the challenge for this training program was to tackle the different perspectives with a process view approach already in the orchestrations of such digital governance implementation efforts. The program made sense for those who found the fun of IT.

In this study it becomes a tempting conclusion to draw that a voluntary model drives sense-making of implementation more effectively than mandatory models that risks to break sense-making in the organization. However, it is not that easy. There are more contextual factors that are embedding the sense-making and breaking processes. Here we have pointed at the educational curriculum, educational format, teachers' discretion and professionalism and local work culture. Our follow up study indicated that school leaders experienced sense-breaking whereas the pre-school leaders continuously experienced sense-making in their ongoing implementation.

This theoretical contribution would have been impossible for us to address if we had not combined the proximity dimension of participation in groups discussions with a follow up study one year after to address the distance dimension to ensure for the critical friend component in on-going evaluation and collaborative research approaches. The strong connection to empirical investigations in theory became evident [45]. It is clearly the case that there is a theoretical need to elaborate on the importance of sense making when implementing digital policies in local governments. This is important since most programs based on policy initiative stems from top-down requirements. Therefore, there is a need to further elaborate on the relation between accountability and sense-making in implementation with a process approach to enable a more robust theoretical understanding of the intricate relations that need attention for future recommendation to practice.

Through the narratives of this case study of a program aiming to train leaders in schools and pre-schools to enhance digital transformation there was a search to make sense of the training in their daily leadership context. The school leaders had to follow the program and for most of them the training became sense-breaking since what they learned in the program did not make sense. On the other hand, the pre-schools leaders that could join the program voluntarily, made sense of the training and found the fun of IT.

References

1. Andersson, A., Hatakka, M., Grönlund, Å., Wiklund, M.: Reclaiming the students–coping with social media in 1:1 schools. Learn. Media Technol. **39**, 37–52 (2014). https://doi.org/10.1080/17439884.2012.756518
2. Andersson, A., Hedström, K., Siegert, S., Sommar, C.: Teachers Falling off the Cliff: Affordances and Constraints of Social Media in School. In: Proceedings of the 54th Hawaii International Conference on System Sciences, pp. 2995–3004 (2021). http://hdl.handle.net/10125/70979
3. Greenhow, C., Lewin, C.: Social media and education: Reconceptualizing the boundaries of formal and informal learning. In: Selwyn, N., Stirling, E. (eds.) Social Media and Education, pp. 6–30. Routledge, London (2019)
4. Bergdahl, N., Nouri, J., Fors, U., Knutsson, O.: Engagement, disengagement and performance when learning with technologies in upper secondary school. Comput. Educ. **149**, 103783 (2020). https://doi.org/10.1016/j.compedu.2019.103783
5. Choi, M.: A concept analysis of digital citizenship for democratic citizenship education in the internet age. Theory Res. Soc. Educ. **44**, 565–607 (2016). https://doi.org/10.1080/00933104.2016.1210549
6. Gunnars, F.: A systematic review of special educational interventions for student attention: executive function and digital technology in primary school. J. Spec. Educ. Technol. **39**, 264–276 (2024). https://doi.org/10.1177/01626434231198226
7. Tondeur, J., Devos, G., Van Houtte, M., van Brakk, J., Valcke, M.: Understanding structural and cultural school characteristics in relation to educational change: the case of ICT integration. Educ. Stud. **35**, 223–235 (2009). https://doi.org/10.1080/03055690902804349
8. Selwyn, N.: 'It's all about standardization': exploring the digital (re)configuration of school management and administration. Camb. J. Educ. **41**, 473–488 (2011). https://doi.org/10.1080/0305764X.2011.625003
9. Hammond, M.: Introducing ICT in schools in England: rationale and consequences. Br. J. Edu. Technol. **45**, 191–201 (2014). https://doi.org/10.1111/bjet.12033
10. Heidlund, M., Gidlund, K.L.: The making of digitalization: like nailing jelly to a wall. Inf. Polity **28**, 29–42 (2023). https://doi.org/10.3233/IP-220007
11. Reisdorf, B., Rhinesmith, C.: Digital inclusion as a core component of social inclusion. Soc. Incl. **8**, 132–137 (2020). https://doi.org/10.17645/si.v8i2.3184
12. van Dijk, J.: The Digital Divide. Wiley, Hoboken (2020)
13. Bernhard, I., Wihlborg, E.: Bringing all clients into the system–professional digital discretion to enhance inclusion when services are automated. Inf. Polity **27**, 373–389 (2022). https://doi.org/10.3233/IP-200268
14. Wihlborg, E., Iacobaeus, H.: Context matters: different entrepreneurial approaches among street-level bureaucrats enhancing digital inclusion. European Policy Analysis. **9**, 379–396 (2023). https://doi.org/10.1002/epa2.1197
15. Bannister, F.: Beyond the box: Reflections on the need for more blue sky thinking in research. Government Information Quarterly. **40** (2023). https://doi.org/10.1016/j.giq.2023.101831
16. Bergviken Rensfeldt, A., Player-Koro, C.: Skolans digitalisering – läroplan och styrning. In: Godhe, A., Sofkova Hashemi, S. (eds.) Digital kompetens för lärare, pp. 45–66. Gleerups, Falkenberg (2019)
17. Velki, T., Miocic, M.: Applying TAM (Technology Acceptance Model) to predict effective ICT use of preschool teachers during the COVID-19 pandemic. 46th MIPRO, pp. 527–532 (2023). https://doi.org/10.23919/MIPRO57284.2023.10159689
18. Selwyn, N.: School and digitalization. Is education better with digital technology? Diadalos, Göteborg (2017)

19. Grönlund, Å.: Changing the school with technology: Beyond one computer per student. Örebro University, Örebro (2014)
20. Schrum, L., Levin, B.B.: Educational technologies and twenty-first century leadership for learning. Int. J. Leadersh. Educ. **19**, 17–39 (2016). https://doi.org/10.1080/13603124.2015.1096078
21. Håkansson Lindqvist, M.: Conditions for Technology Enhanced Learning and Educational Change: A case study of a 1:1 initiative. Doctoral dissertation, Umeå University (2015)
22. Vanderlinde, R., van Braak, J.: The e-capacity of primary schools: development of a conceptual model and scale construction from a school improvement perspective. Comput. Educ. **55**, 541–553 (2020). https://doi.org/10.1016/j.compedu.2010.02.016
23. Balanskat, A., Bannister, D., Hertz, B., Sigillò, E., Vuorikari, R.: Overview and analysis of 1:1 learning initiatives in Europe. Publications Office of the European Union, Brussels (2013). https://doi.org/10.2791/20333
24. Blossing, U., Liljenberg, M.: School leaders' relational and management work orientation. Int. J. Educ. Manage. **33**, 276–286 (2019). https://doi.org/10.1108/IJEM-07-2017-0185
25. Håkansson Lindqvist, M., Pettersson, F.: Digitalization and school leadership: on the complexity of leading for digitalization in school. Int. J. Inf. Learn. Technol. **36**, 218–230 (2019). https://doi.org/10.1108/IJILT-11-2018-0126
26. Lim, N., Grönlund, Å., Andersson, A.: Cloud computing: the beliefs and perceptions of Swedish school principals. Comput. Educ. **84**, 90–100 (2015). https://doi.org/10.1016/j.compedu.2015.01.009
27. Henrekson, M., Wennström, J.: Dumbing Down: The Crisis of Quality and Equity in a Once-Great School System—And How to Reverse the Trend. Springer, Cham (2022).https://doi.org/10.1007/978-3-030-93429-3
28. Mårell-Olsson, E., Bergström, P.: Digital transformation in Swedish schools: Principals' strategic leadership and organisation of tablet-based one-to-one computing initiatives. Seminar.net: Media, technology and lifelong learning. **14**, 174–187 (2018). http://urn.kb.se/resolve?urn=urn:nbn:se:umu:diva-152770
29. Gustafsson, U.: Taking a step back for a leap forward: policy formation for the digitalisation of schools from the views of Swedish national policymakers. Educ. Inq. **12**, 329–346 (2021). https://doi.org/10.1080/20004508.2021.1917487
30. Wibeck, V., Linnér, B.O.: Sense-making analysis: a framework for multi-strategy and cross-country research. Int J Qual Methods **20**, 1–12 (2021). https://doi.org/10.1177/16094069211998907
31. Spante, M.: Leda digitalisering. Working paper 20200214, University West (2020)
32. Nilsson, L., Sorbring, E.: Samverkansforskning: att främja barns och ungas välfärd. Liber, Stockholm (2019)
33. van de Ven, A.: Engaged Scholarship: A Guide for Organizational and Social Re-search. Oxford University Press, Oxford (2007)
34. Gunnarsson, E., Hansen, H.P., Nielsen, B.S., Sriskandarajah, N.: Action research for democracy: New Ideas and Perspectives from Scandinavia. Routledge, London (2015). https://doi.org/10.4324/9781315659909
35. Albinsson, G., Arnesson, K.: How critical can you be as an on-going evaluator? Int. J. Action Res. **6**, 256–287 (2010). https://nbn-resolving.org/urn:nbn:de:0168-ssoar-414115
36. Weick, K.: Sensemaking in Organizations. Sage, Thousand Oaks (1995)
37. Weick, K., Sutcliffe, K.M., Obstfeld, D.: Organizing and the process of sensemaking. Organ. Sci. **16**, 409–421 (2005). https://doi.org/10.1287/orsc.1050.0133
38. Foreman-Wernet, L., Dervin, B.: Hidden depths and everyday secrets: how audience sense-making can inform arts policy and practice. J. Arts Manage. Law Soc. **47**, 47–63 (2017). https://doi.org/10.1080/10632921.2016.1229642

39. Coburn, C.E.: Shaping teacher sensemaking: school leaders and the enactment of reading policy. Educ. Policy **19**, 476–509 (2005). https://doi.org/10.1177/0895904805276143
40. Pyhältö, K., Pietarinen, J., Soini, T.: Teachers' professional agency and learning: From adaption to active modification in the teacher community. Teachers Teach. **21**, 811–830 (2015). https://doi.org/10.1080/13540602.2014.995483
41. Nordholm, D.: State policy directives and middle-tier translation in a Swedish example. J. Educ. Adm. **54**, 393–408 (2016). https://doi.org/10.1108/JEA-05-2015-0036
42. Wilson, C., Mergel, I.: Overcoming barriers to digital government: mapping the strategies of digital champions. Gov. Inf. Q. **39**, 101681 (2022). https://doi.org/10.1016/j.giq.2022.101681
43. Cornelissen, J.P.: Sensemaking under pressure: the influence of professional roles and social accountability on the creation of sense. Organ. Sci. **23**, 118–137 (2012). https://doi.org/10.1287/orsc.1100.0640
44. Ybema, S., Willems, T.: Making sense of sense-breaking. Acad. Manage. Proc. 14563 (2015). https://doi.org/10.5465/ambpp.2015.14563
45. Alvesson, M., Kärreman, D.: Constructing mystery: empirical matters in theory development. Acad. Manage. Rev. **32**, 1265–1281 (2007). https://doi.org/10.5465/amr.2007.26586822

E-Participation Without Democracy: Understanding Variation in Digital Engagement in Non-democracies

Thomas Hayes and Martin Karlsson

Örebro University, Fakultetsgatan 1, 701 82 Örebro, Sweden
{Thomas.hayes,martin.karlsson}@oru.se

Abstract. The variation in E-participation adoption and obstruction among non-democratic regimes is not sufficiently understood in earlier research. We attribute this to a lack of conceptual instruments for systematically studying the regime attributes of non-democratic states. Inspired by the work of Linz and Stepan (1), we demonstrate how a more fine-grained and multi-dimensional taxonomy of non-democratic regimes could differentiate between regime behaviours. Based on this categorisation, we further formulate expectations regarding four dimensions of E-participation in different types of non-democratic regimes. We argue that the proposed regime categorisation and identified expectations can form a basis for more nuanced comparative research on E-participation in non-democratic states.

Keywords: E-participation · Non-democratic regimes · Comparative politics · regime categorisation

1 Introduction

Many non-democratic states engage in and even outperform democracies in terms of E-participation [2–5] while retaining (and even strengthening) non-democratic rule [6]. This puzzling empirical reality has inspired a wave of research devoted to deciphering the mechanisms of e-participation in non-democratic states [c.f 7, 8]. However, such studies have given limited attention to distinguishing between differing flavours of non-democratic regimes, leading to a staid and limited assessment of the relationship between regime characteristics and aspects of E-participation. As Kabanov [9, p. 2] argues, "[…] while being focused on the delineation between authoritarian and democratic countries, they overlook different dynamics within both types".

While focused comparisons and case study research have observed varying dynamics of how different non-democratic regimes have approached ICTs in general and E-participation in particular [cf. 10], this knowledge has, to a limited extent, transferred to global comparative studies of E-participation.

The objective of the present study is to contribute to developing a more nuanced and valid understanding of the varying dynamics of e-participation in non-democratic regimes. This objective is addressed through four research tasks: (1) reviewing and discussing regime classifications in earlier research, (2) empirically investigating the extent to which country-level measurements of E-participation vary across non-democratic states, (3) inspired by the work of Linz and Stepan [1] demonstrate the application of a more fine-grained taxonomy of non-democratic regimes, and (4) based on this categorisation demonstrate differing expectations regarding e-participation in different types of non-democratic regimes for future research.

1.1 Methodology

Following Karlsson and Adenskog [11], we define e-participation as encompassing processes of civic engagement in politics supported by ICTs. However, in the context of this study, we focus on a subset of E-participation processes that Kersting [12] calls "invited spaces", that is, processes of ICT-enabled civic participation that are initiated by public institutions. It is concerning such processes that regime characteristics become most relevant. Operationally, we utilise the UN's E-Government Survey [13] E-participation index (EPI) as a measure of E-participation in different countries. Further, in formulating expectations regarding how E-participation is enacted in different types of non-democratic regimes we focus on the three dimensions this index: (1) E-information: Enabling participation by providing citizens with public information and access to information without or upon demand; (2) E-consultation: Engaging citizens in contributions to and deliberation on public policies and services; (3) E-decision-making: Empowering citizens through co-design of policy options and co-production of service components and delivery modalities. It should be noted that while the index from which these definitions are drawn offer a comprehensive source of data on E-participation in the countries of the world over 20 years, the EPI has some clear insufficiencies. First, the EPI has been widely criticised for disregarding the political context of countries, resulting in a suspected overvaluation of E-participation in non-democratic countries [cf. 2, 3, 9, 14–16]. Second, the EPI is criticised for applying a low bar regarding what suffices participation, including E-information as an aspect of E-participation [17]. However, despite these insufficiencies the EPI still constitutes the best available data source for global comparisons of E-participation. While promising alternatives have been developed, they are yet to be implemented at the greater scale that the analysis of this paper requires [18].

Moving forward, this study's central underlying assumption is that the political regime's characteristics are consequential for if and how electronic participation is adopted in a country. There are two primary reasons to make such an assumption. First, substantial research has demonstrated that regimes adopt new technologies differently [19–21]. As North [19] argues, states can through legislation, regulation, taxation and subsidies, hinder or support the spread of new technologies in a country. What is more, different types of regimes may have differing interests in a specific technological change, as it tends to create new winners and new losers [20, p.183). Second, electronic participation is not only a question of technology adoption. Rather, e-participation entails the creation of channels and processes for citizens to participate in politics. While generally thought of as intrinsically linked to democratic regimes, research on "democratic authoritarianism" [cf. 22] has demonstrated that non-democratic regimes perform essential functions related to the regime's survival by initiating citizen participation processes. However, such functions of citizen participation tend to vary between different types of regimes [23].

In order to investigate the above, the remainder of the paper is structured as follows. In the next section, we review large N comparative studies of E-participation in different countries, focusing on what regime classifications are used to differentiate among democratic regimes. Thereafter we present statistical analyses of the variation in E-participation performance among non-democratic regimes and the extent to which this variation can be explained using the most common classifications in earlier research. We then turn to our theoretical contribution to the field. First, we present the regime classification developed by Linz and Stefan [1]. Thereafter we apply this classification to the area of E-participation with the aim to identify expectations in terms of enactment and obstruction of E-participation in different types of non-democratic regimes. In the last section, we present the main conclusions of the paper.

2 Regime Categorisations in Earlier Research

Most earlier studies of e-participation in non-democratic regimes treat such regimes in one of two ways (Table 1). Non-democracies are lumped together in one group, distinguished from democratic regimes [cf. 3, 4], or non-democratic regimes are categorised in their degree of democratisation [cf. 24, 25], for instance, in terms of degrees of freedom (partly free and not free) or as hard- and soft authoritarian regimes.

For both these types of studies, the focus is primarily on understanding differences and similarities in e-participation development between democratic and non-democratic, or degrees of democratised states. The primary theoretical basis for such categorisations is a democratisation theory of internet governance [26, 27], postulating that ICTs affordances of greater government transparency, efficient mass communication and that coordination of civic action poses opportunities for the reinvigoration of democratic systems of government and threats to the regime survival of non-democratic regimes. These categorisations imply that the one characteristic necessary to understand non-democratic regimes is visible in their label: they are not democracies. Further, such categorisations also imply that the trajectory of regime change follows the axis between democratic and non-democratic, meaning that regimes either move towards being democratic (democratisation) or non-democratic (deconsolidation). However, earlier research has shown that only 25% of regime changes have led to democratisation [28]. Most regime changes led to transitions from one form of non-democracy to another, indicating the non-democratic category as highly heterogeneous and needing further differentiation.

Only two studies of e-participation in non-democracies have moved past these one-dimensional categorisations and created more nuanced taxonomies of non-democratic regimes [27, 29]. The taxonomies in these studies share strong similarities as they focus on regime leadership. Both these taxonomies differentiate between civil and military regimes. Further, among civil regimes, monarchic and personalist regimes are distinguished from party-based regimes. Lastly, these studies also differentiate between party-based regimes based on the level of fragmentation of the party system, i.e. if a single party or multiple parties are allowed to operate in the country.

These more diverse regime typologies that focus upon leadership as a lens for regime classification expand the diversity of regime categorisation on a relative scale. However, the conceptualisation of only a singular new dimension in the guise of leadership still presents a particularly broad and underdeveloped stroke at regime classification of non-democracies. Leadership is a vital area of interest within regime classification and a central node in a state's levers of power; however, what is critical is that it does not exist in a vacuum. Reich [30, p. 5] notes that the focus on leadership limits the work's application in political regime classification as, in reality, regimes comprise more than the rules and indicators of who leaders are and how they are chosen. Thus, the resulting studies provide a narrow focus that sacrifices conceptual validity for reliability. This assessment is corroborated by Kabanov [9], who finds that within his study on aspects of legitimation, via this dimension of regime categorisation, such processes cannot be explained by regime taxonomy alone.

Table 1. Taxonomies of non-democratic regimes in comparative studies of e-participation.

Article	Categorization of non-democratic regimes				
	1	2	3	4	5
Åström et al., [3]	Non-democracy				
Karlsson, [4]	Non-democracy				
Cho & Rethemeyer, [31]	Autocracy				
Katz & Halperin, [24]	Partly free	Not free			
Woo & Kübler, [24]	Soft authoritarian	Hard authoritarian			
Choi & Yee, [32]	Electoral authoritarian	Authoritarian			
Kneuer & Harnisch, [27]	Deficient democracy	Military	Multiparty	Single party	Monarchy
Kabanov, [29]	Monarchy	Military	Multiparty	Single party	No party

2.1 Variation in Internet Governance Measures Among Democracies and Non-democracies

One potential reason for handling non-democratic countries as a homogeneous group sharing hypotheses and mechanisms would be that there is a slight variation in e-participation within this group of countries. If there are little differences to explain, there is also little need for differentiation between types and sub-types of non-democratic countries. However, comparative studies have repeatedly found that the level of democratisation is a poor predictor of E-participation [3, 9, 14–16]. Our analyses (Table 2 below) verify this result, indicating that a simple dichotomous categorisation into democracies and non-democracies (based on the QoG-dataset) [33] explains only a small share of the variation scores on the UN e-participation index among countries. During the period 2003–2020, the share of variance in e-participation explained by a dichotomous division between democracies and non-democracies is between 2.9 and 18.2% (Eta^2: .029–.182), which means that between 81.8 and 97.1% of the variation is within these categories. Continuous measurement of democracy (Polity V) [34] performs slightly better in explaining between-country variation in e-participation during the same period between 4.9 and 28% of the variation is explained by the level of democracy (R^2: .049–.28). However, this one-dimension of regime differences does not explain the lion share of variation in e-participation between countries.

Further, the variation among non-democratic countries has increased substantially during the time period. Table 2 below presents the variance among non-democracies in the UN E-participation index (2003–2020) [13]. The variance of a distribution measures how much individual data points deviate from the mean, indicating the extent of variation in the data. As the table (leftmost column) shows, the variance steadily increases over time and is almost twice as large in 2020 as in 2003 (σ^2: .025–.043).

Therefore, our analyses indicate that non-democratic countries are becoming more heterogeneous regarding E-participation, and neither a dichotomous nor a continuous measure of democratisation provides a satisfactory explanation for this heterogeneity. We, therefore, suggest that comparative E-participation research should explore more fine-grained theoretical tools for understanding and categorising non-democratic regimes. In doing so, we further recommend that researchers in the field of E-participation should draw inspiration and knowledge from the field of comparative politics. In the following sections, we will present a potential taxonomy of non-democratic regimes and formulate possible expectations on E-participation for each type of non-democratic regime.

Table 2. Variance in the UN EPI among non-democratic regimes and the share of variance explained by dichotomous and continuous measures of democracy, 2003–2020.

	Variance among non-democracies	Share of variance explained	
	(using dichotomous categorisation)	*Dichotomous categorisation*	*Continuous democracy index*
2003	.025	.182	.280
2004	.025	.147	.225
2005	.029	.136	.199
2008	.020	.048	.082
2010	.021	.059	.114

(continued)

Table 2. (*continued*)

	Variance among non-democracies (using dichotomous categorisation)	Share of variance explained	
		Dichotomous categorisation	*Continuous democracy index*
2012	.032	.029	.049
2014	.036	.066	.087
2016	.037	.114	.110
2018	.039	.124	.116
2020	.043	.135	.137

3 Regime Classification

The classification of regimes constitutes a fundamental aspect of comparative political science. As Alvarez and colleagues [35, p. 3] argue, "To study systematically issues concerning both the origins and the consequences of political regimes, we need valid and reliable classifications". Today, the amount of available classifications and datasets measuring different aspects of regimes is abundant [36, p. 531]. This easy access to the building blocks of large-N comparative regimes studies has instigated exponential growth. However, it has also made it easy to skip the vital step of carefully considering and theorising around the potential mechanisms and expectations regarding the conduct and actions of different regimes.

The following section will present a potential classification of non-democratic regimes to analyse e-participation among countries worldwide. This classification is based on the work of Linz and Stepan [1]. Its use rests on the assumption that non-democratic regimes can be qualitatively divergent on several dimensions apart from their varying levels of "democraticness" [37, p. 7]. In other words, regimes can be steadily non-democratic but still substantially different from each other, and these differences can be heavily consequential for how the regime operates (for instance, concerning E-participation). Such an assertion is supported by Wright [38], who argues that within non-democracy, the differing relationships between critical institutions such as the military, the party and the leadership are often orthogonal to measures of democracy.

Not content with a simple delineation between democracy and autocracy, Linz and Stepan [1] propose five different categories of regimes: Democratic, Totalitarian, Post-Totalitarian, Sultanistic, and Authoritarian regimes. This taxonomy is undoubtedly an expansion to the previous di/trichotomous approaches to classification; however, it remains restrictive enough that each regime category still encompasses groups of regimes that can be compared internally and externally against other categories. To be clear, this taxonomy is not the solution to the issue that this paper tackles; its age is apparent in the typologies that it produces, some of which (totalitarian and post-totalitarian) are not suitable for the contemporary setting. Instead, the critical principles of the taxonomy and what this paper argues should be incorporated within future research and a new taxonomy. Therefore, the taxonomy proposed by Linz and Stepan [1] is a means to an end, merely a demonstration of the different expectations that a more nuanced approach could deliver.

In order to accomplish this, what sets Linz's [1] work apart is that the taxonomy builds on a multidimensional approach to regime categorisation. The five regime types differ along four primary indicator areas: pluralism, mobilisation, leadership, and ideology. This approach further allows for a potential spectrum of sub-classifications utilising these four indicator dimensions. This would create a potentially multi-level and multidimensional approach within which non-democracies can be compared collectively against democracies (level 1), can be compared independently against democracy and each other (level 2), and finally can be compared internally within one of Linz typologies (lvl 3). This demonstrates the potential for analysing different levels of abstraction, which, as Munck and Verkuilen [39] note, is critical to avoid either an overly minimalist or maximalist approach to conceptualisation. This gives greater scope for its application to a broader semantic field of regime application, especially within the context of non-democratic regimes, which can be as different from each other as they are from democracies [40]. Moreover, using such dimensions allows Linz and Stepan [1] to capture institutional and non-institutional measures of non-democracy, which, as Wright [38] notes, is critical to providing systematic information on how such regimes work and differ from each other.

Something that is also of particular note is that this multidimensional approach allows for the conceptual space to move away from the previous consensus of a single linear spectrum-based assessment of regime type, within which there is a singular direction of travel from autocracy to democracy. Instead, as mentioned previously, each of Linz and Stepan's [1] regime categorisations acts as a start and end point of regime transition, denoting particular regime changes and behaviours and allowing multiple travel directions regarding regime type at any given moment. By doing this, the taxonomy removes the critical assumption of inherent democratisation of regimes over time [cf. 41, 42]. In a time when democracy is arguably in a global recession [43] and in the face of evidence that most regime transitions are between different types of non-democratic regimes [28], this assumption is on shaky ground.

3.1 Authoritarianism

Linz and Stepan [1, p. 45] define authoritarianism as a specific type of regime with distinct features rather than seeing it as a synonym for non-democratic regimes in general. Within their conceptualisation, an authoritarian regime is a political system with limited and non-responsible political pluralism, meaning that there are limits on oppositional political actors' (e.g. elected representatives, political parties and interest groups) possibilities to operate and absolute limits on their possibilities to partake in the responsibility of steering the state. Likewise Purcell [44, p. 303; 1] notes within the context of the Mexican government of the 1970s for example, that within the above specific interest groups exist, however, with limited autonomy. Such entities are best understood as adjuncts of the central authorities, implementers of the regime leadership decisions, rather than defenders of their rank-and-file members. These groups act unhindered as long as they do not impact the authority of the authoritarian leader who rules over them [1]. Interestingly, such entities also have bureaucratic structures and norms that limit what the authoritarian leaders can ask them to do. Often, the authoritarian leader's power is secured by placating or balancing the power of such interest groups, especially within the context of societal elites, who have the potential to change the balance of power within the state. Conversely, social and economic pluralism is extensive as the market can operate more freely, and the state's interference in social life and organisation is limited, creating an overall context within which the ruler lacks total control [1].

Within the typology, an authoritarian regime lacks a state ideology; instead, it is composed of distinct mentalities, such as a commitment to nationalistic ideals. An authoritarian state would further lack any significant mobilisation apart from particular points in its development, such as an initial coup to gain control over the state. Thus, there would be both low regime mobilisation and low citizen participation, with political inclusion limited to the cooptation of the state's elite groups to secure the tenuous grip that authoritarian leaders have on power [1]. Within this typology, the influence of the nature of overarching authority and order is visible, but only within a specific dimension, within which, as outlined above, a single person or small group acts as the authority of the state via a high degree of personalism, with the order of the state structures aligned to benefit the ruling individual [1].

Depending upon the levels and significance of pluralism, especially within the state's elite class, one would expect varying low levels of E-participation, almost as a function of patronage, to sustain the authoritarian dictator's power [45]. This would largely depend upon the leadership type and subsequent supporting power structures that the autocrat ruling a regime relies upon [46]. Available public information would be minimal, with access being a two-way street, in which public sites and internet control are used to surveil potential dissidents, a single tool within a more extensive reactive control apparatus. E-participation would be expected to become an exclusive activity, and while presented as open to all, it would disproportionately be beneficial to balance the power-seeking behaviours of political rivals rather than as a function of political inclusion [47]. Political pluralism within such E-participatory forums would exist but be tolerated within arbitrary bounds and controls determined by the government, with harsh reprisal for those who overstep the mark. Overall, participation is not substantially encouraged, nor is a substantial mobilisation cultivated outside of mentalities beneficial to the regime ruler, such as xenophobia [48]. Mobilisation would be expected to be reduced via restrictive control over access to the technologies required for E-participation to actively repress alternate views. It is within this that the pseudo-nature of E-participation is drawn. Authoritarian regimes would not be expected to pursue participation or mobilisation; thus, e-participation is only a democratic facade. Its function is one of pure autocratic interest as a tool to placate the international community and to mitigate the lack of total control that such leaders have over their regimes [7, 48].

3.2 Totalitarianism

Continuing this regime taxonomy with totalitarianism, which, as Linz and Stepan [1, p. 45] define as an autocratic regime within which almost all pre-existing economic, political and civil pluralism is eliminated and replaced by a well-articulated utopian ideology and a total monopoly over power by the state and the party. Denizens of such a state derive their drive and mission via a holistic conception of this utopianistic worldview. They are genuinely committed individuals, even if the ideology is forcefully disseminated to the regime's denizens via a robust, overbearing state apparatus. Thus, this ideology is propped up by substantial support expressed via high mobilisation for the regime specifically and high levels of obligatory citizen participation within state structures. Such a regime is guided by top-party leadership that rules charismatically with undefined limits to their power, creating a context with extensive unpredictability and vulnerability for citizens of the regime. Candidates for the leadership are drawn from the party's upper echelon and chosen based on their success and commitment to either the party or the ideology. This is not to say that such candidates are not voted upon, but that all candidates are drawn from the same and only political party on the ballot, creating a facade of choice for most citizens [1].

In a totalitarian regime, incorporating E-participation would have the primary goal of maintaining control and managing public perception alongside fostering genuine citizen engagement or empowerment, which would generate high participation and mobilisation. However, such a regime would also look to simultaneously reduce levels of pluralism within E-participation processes. This could be demonstrated, for example, via Kang's [49] assessment of administrative adsorption within what they identify as the process in which neo-totalitarian China makes concerted efforts to make sure digital civil society and public spaces cannot emerge outside the regime through administrative and regulatory policy while developing an alternate state-led civil-society. It would be expected via Linz and Stepan's [1] analysis that such a state would have high levels of political propaganda within participation channels, with such sites having a dual role, further allowing the state to monitor access and identify dissenters within the regime. As Kang [49] further notes [...] "In the social realm, China's neo-totalitarian regime constrains the people's freedom of speech, assembly, and association and regulates social organisations via the governing system of AAS." As such, one would expect that participation channels would have significant control over information access via online filtering and localised control over the information presented internally and externally, directing participation towards pro-regime sentiment, with no allowed presence of alternative narratives. There would be the development of pseudo-participation forums. However, the restrictions of pluralism required for totalitarianism would mean that such forums would only allow for participation in the narrative and ideology of the state rather than fostering honest political debate and dialogue. There would be the potential for E-decision-making. However, such avenues would be restricted to low-level decisions, which would further be highly redundant as the lack of plurality means that the results of such forums would act as an illusion of choice.

3.3 Post-totalitarianism

When a totalitarian regime falls, Linz and Stepan argue that some enduring legacies of the totalitarian era set post-totalitarian regimes apart from other non-democratic typologies. Linz and Stepan [1, p. 45] identify a post-totalitarian regime as having limited social and economic pluralism, with no political pluralism due to the overbearing presence of the state within this dimension. The state sustains a weakened monopoly over all forms of power. Most plurality is expressed through tolerated state structures and dissident groups that exist in a "second society", which opens the regime up to new dynamics of vulnerability. The ideology of the totalitarian regime remains, but again, within a weaker state, within which the former utopian ideology is replaced by a programmatic consensus that exists with limited debate and the assertion of rational decision-making outside of the ideological framework. It is often found within such regimes that the former state ideology and its claims clash with reality, leading to a greater focus on performance-based rather than ideological measures of success [1, p. 49]. Mobilisation by the state is only to achieve the minimum required levels of compliance and conformity, with a progressive loss of interest in generating political participation outside such dimensions. Those mobilised are often no longer true believers but careerists and opportunists participating for personal gain. The leadership of such a regime, like its predecessor, is party-based; however, it is defined by its focus on personal security. Its figureheads are drawn from the political party, without necessarily through displays of commitment and success to the state/party. The leadership is more bureaucratic and less charismatic than its totalitarian alternative, with significantly less power invested in the top leadership, reducing the high levels of unpredictability for elites and citizens alike.

Within E-participation and the proactive modalities of steering the population, this would be reflected in less ideologically influenced information channels, with propaganda on public forums reflecting a need to generate a consensus upon rational and positive performance-based assessment of the regime rather than espouse a utopian platform. Public Information would still be provided on a more restricted basis, with negative assessments of the regime being hidden from sight, and information access would remain a two-way street to identify dissidents within the state. Like totalitarianism, there would still be high levels of obligatory participation via government consultation channels; however, it would only be enough to sustain necessary support for the regime as a means to an end rather than the desire for participation and mobilisation as the end itself. In turn, this would be reflected by increasing levels of apathy and disengagement from such political processes as the privatisation of citizen values increases. This would be reflected within a reverse process of administrative adsorption, [49], which is almost an administrative regression within which the weakening of ideology leads to a lack of cohesion needed for complete totalitarian control over digital structures. Within this consultation, forums would be able to have limited forms of social and economic pluralism expressed at levels acceptable to the state. Some opposition mobilisation may occur from these other forms of plurality, creating potential political opposition to the regime.

3.4 Sultanism

The final non-democratic regime type that Linz and Stepan [1, p. 51] identify is Sultanistic regimes, which have strong tendencies towards patrimonialism. As such, the administration and its instruments are the personal tools of a single despot. There is a high fusion of the public and private spheres by the ruler of the given regime, within which the polity is entirely their domain. While various forms of civil and economic pluralism exist, political pluralism is non-existent, especially in forms that would divest power from the regime ruler. Due to the weak institutionalisation and lack of the rule of law, all forms of mobilisation, ideology and pluralism are subjected to random and unpredictable intervention by the regime ruler, which makes any form of power other than the Despot's precarious [1, p. 53]. Sultanism, like totalitarianism, utilises ideology to a significant degree. However, this is not an elaborate exercise in utopian guidance or coercion but rather a highly personalistic approach to an ideology co-opted by the despot for their benefit and moral justification. The Sultanistic regime only requires the mobilisation of forces that the ruler can wield to secure their absolute power, with such mobilisation reinforced by solid ties of clientelism and the presence of this faux personalistic ideology that encourages fanaticism. Within Totalitarianism, ideology is arguably the only constraint upon the actions of the leadership, while sultanistic regimes have no such limitations, making them the expression of unrestrained personal rulership, and is a critical variable in differentiating the forms of absolute power that are present in both.

In a sultanistic regime, E-participation policies would likely be strategically implemented to maintain the regime's authority and glorify the personalistic ruler's benevolence in highly arbitrary and abstract manners. Thus, the regime would be expected to engage selectively in controlled online forums and platforms, using them to disseminate symbolistic propaganda and regime-approved information. While there would be a mix of public and private access channels to public information, all would be under the thumb of the sultan, and access would be a two-way street, giving the regime the capability to identify and persecute dissidents. Such information access would be minimal and highly restricted to prevent mobilisation against the regime and disrupt the narrative of the benevolence of the regime's ruler. This is observable within the current Saudi Arabian regime for example, within which crackdowns on dissidents and clerics who think King Abdullah's reforms are either too permissive or need to go further alike [50]. Within this, what would, at the surface level, be considered proactive measures to steer the population's interaction with information would be reactive measures of control. This would be demonstrated by the propaganda within the E-information channels available, which would be connected to a guiding ideology as with totalitarian regimes; however, the purpose of it is not to mobilise the entire population in genuine support for the regime or its utopian future, instead to mobilise fanatic para-state groups who can and will act at the behest of the regime ruler in reactive repression of the regimes population. This may occur within soft and complex forms of power; as Wehrey [50] notes, in Saudi Arabia, social media and government pages upon which citizens can interact are often utilised to channel and mobilise Twitter armies of regime loyalists to push the regime's ideological messages. In turn, social media interactions that criticise Islam or the regime rulers have resulted in prison sentences, demonstrating the regime's synergy in applying both methods to repress the population.

Moving on, there would be a shallow level of e-consultation only to present a positive external image to the outside world via pseudo-participation in a fashion similar to that of an authoritarian regime [7, 48]. However, such a facade would not encompass power-sharing. E-consultations would have relatively low direct control via law and regulation; however, punitive measures would be taken against those who criticise the regime or challenge the right of the Sultan to their power.

3.5 Summary

As shown in Table 3 (above), some differences are outside the assumed dimension of democratisation. For example, the differentiation in the forms of propaganda between the regimes within the e-information modality of participation is particularly apparent. Even when regimes both are expected to utilise ideological forms of propaganda, such as totalitarian and sultanistic regimes, within this taxonomy, the forms of ideology would be distinguishable, and the relationship with levels of access would interact differently. Likewise, the levels of expected access within the e-consultation sphere within totalitarian and post-totalitarian regimes would separate them from authoritarian and sultanistic regimes. However, such regimes could then be differentiated from each other by levels of engagement and plurality. These dimensions conform to the minimalist approach of grouping non-democracies, which would, at worst, miss entirely and, at best, be unable to explain.

Table 3. Expectations regarding enactment of E-information, E-consultation and E-decision making in divergent types of non-democratic regimes.

Regime Type	E-Information	E-Consultation	E-decision making
Authoritarian	The use of information access as a two-way street to actively repress dissidents Low-medium levels of access, high levels of control over information Lack of ideological propaganda channels	Defining feature is the use of consultation processes particularly for power-balancing/sharing acts Limited plurality in E–consultation Low levels of participation, Highly controlled	Involvement in decision making is likely to only be a facade to convince an international audience
Totalitarian	The use of information access as a two-way street to actively repress dissidents High use of ideological utopian propaganda within E-information channels High levels of access, high levels of control	High levels of citizen engagement in E-consultation No plurality in E-consultation E-consultation allowed only in the limits of the guiding ideology Universal Obligatory Participation in E-consultation channels	Potential for E-decision-making forums but restricted to low-level decisions with limited impact Lack of plurality renders the results of such forums as illusory choices, contributing to the regime's control narrative
Post-totalitarian	Emphasis on less ideologically influenced information channels Performance based propaganda in information channels Two-way information access used for identifying dissidents High levels of access, high levels of control	Limited obligatory participation in government consultation channels Limited social and economic pluralism in E-consultation Middling levels of participation Medium - high levels of control	The focus would appear to be more on control mechanisms and restriction rather than a participatory decision-making process. Potential for low-level decision making forums that have limited impact or allowed plurality, but develop national consensus
Sultanistic	Highly symbolistic ideological propaganda in information channels Access is a two way street used for identifying dissidents Low access, highly controlled	Limited and likely aimed at exclusively presenting a positive external image Low levels of participation, high levels of control	No dimension for E-decision making, as all power lies within the Despots hands or from within state Bureaucracy controlled by them

4 Conclusions

This study addresses the need for more nuance in conceptualising and classifying non-democratic in comparative research on E-participation. We argue that earlier research has failed to explain the variation in how E-participation measures are enforced and obstructed in divergent non-democratic regimes. We attribute this partly to a lack of conceptual instruments to systematically study the regime attributes of non-democratic states.

The majority of earlier studies that we have reviewed have either lumped non-democratic regimes together in a single group or categorised these regimes purely along a dimension of level of "democraticness" [37, p. 7], meaning how distant the characteristics of those regimes are from living up to principles of democracy. These ways of classifying non-democratic regimes have apparent drawbacks. Our study and earlier research have shown that the level of democracy is a poor explanation of between-country differences in E-participation. Further, such one-dimensional conceptualisations fail to consider significant differences in how non-democratic regimes function and operate.

As an alternative, we have utilised a regime classification of non-democracies inspired by the work of Linz and Stepan [1] that distinguishes between four different types of non-democracies: authoritarian, totalitarian, post-totalitarian and sultanistic regimes. We have summarised the characteristics of these regime types along four dimensions: pluralism, mobilisation, leadership and ideology. Based on these characterisations, we have presented expectations regarding how these regimes would enact E-participation. This exercise shows us that divergent expectations can be drawn from the differences between these regimes regarding how they enact E-participation.

The typology used in this paper has various shortcomings. The collapse of utopianistic ideology within state structures has meant that both totalitarian and post-totalitarian regimes have become relevant categories today than in the 1970s and '80s'. Thus, using this taxonomy precisely as it was conceptualised by Linz and Stepan [1] in reality might lead to an oversaturation of the number of regimes within the authoritarian and sultanistic typologies, leading the researchers back to square one as these categories would no longer explain the levels of heterogeneity between the regimes that fall into their parameters. Moreover, outlying hard-bordered ideal categories do not accommodate the existence of hybridised regimes between the typologies presented above. Some regimes may be in active transition, but others may be stable and sustain themselves over time. Thus, it would be more appropriate to pin the benefits of such a taxonomy via Linz and Stepan's [1] multi-dimensional approach to the development of categorisation alongside the principle of breaking down non-democracies into differentiated regimes. The taxonomy, upon this basis, could be updated and refined for the modern age, utilising potentially the same indicator areas but different or more regime typologies. While such a multi-dimensional taxonomy has potential to support a better understanding of how e-participation is enacted and obstructed in different countries, it is important to take into account that institutional or regime-related aspects are far from the only factors that influences E-participation. For instance, levels of socio-economic development and reliance on foreign trade relations have proven to play a significant role in shaping E-participation

[3]. Further, recent studies warn against investigating E-participation concerning technological, social or institutional factors in isolation and suggest an integrated approach to understanding country-level variations in E-participation [51].

The logical next step is to expand upon this assertion presented within the paper while also addressing the shortcomings of the softer approach to empirical analysis undertaken. This paper has only tested the applicability of dichotomous and continuous categorisations of regimes within the QOG index, making it challenging to apply to alternate indexes such as Freedom House or V-dem that could produce different scores for any given democratic and non-democratic regimes. Such indexes may utilise differing, more expansive or even restrictive operational definitions of democracy and, thus, non-democracy, allowing them to explain variances within e-participation levels to a greater or lesser degree. Therefore, expanding the empirical scope to an alternate index only reinforces the paper's argument. Moreover, it would be a relatively simple task to test the differing typologies in Table 1 if one could test them within the context of the index to which they were applied. This could reinforce the paper's criticism of only utilising one dimension to differentiate between democratic and non-democratic regimes to a greater degree. During this subsequent analysis, one could factor in several control variables for third variables, such as economic development and levels of digitalisation, enriching the understanding of the levels of variance that any given regime typology can explain and further assessing the relevance of including such variables within the typologies themselves. Once the above is achieved, future research could engage in case comparative research, using qualitative assessments of E-participation characteristics in countries and qualitative comparative analysis of differences between countries to develop a new typology to be applied. Such methods might be more helpful in understanding variations in how E-participation is enacted in different non-democratic regimes. Once this typology is established, one could test it for its capacity to explain variance, demonstrating its applicability.

References

1. Linz, J.J., Stepan, A.C.: Problems of Democratic Transition and Consolidation: Southern Europe, South America, and Post-communist Europe. Johns Hopkins University Press, Baltimore (1996)
2. Grönlund, Å.: Connecting eGovernment to real government - the failure of the UN eParticipation index. In: Janssen, M., Scholl, H.J., Wimmer, M.A., Tan, Yh. (eds.) EGOV 2011. LNCS, vol. 6846, pp. 26–37. Springer, Heidelberg (2011). https://doi.org/10.1007/978-3-642-22878-0_3
3. Åström, J., Karlsson, M., Linde, J., Pirannejad, A.: Understanding the rise of e-participation in non-democracies: domestic and international factors. Gov. Inf. Q. **29**(2), 142–150 (2012)
4. Karlsson, M.: Carrots and sticks: internet governance in non–democratic regimes. Int. J. Electron. Gov. **6**(3), 179–186 (2013)
5. Girish, J.W.C.B., Williams, C.B., Yates, D.J.: Predictors of on-line services and e-participation: a cross-national comparison. Gov. Inf. Q. **31**(4), 526–533 (2014)
6. Linde, J., Karlsson, M.: The dictator's new clothes: the relationship between e-participation and quality of government in non-democratic regimes. Int. J. Public Adm. **36**(4), 269–281 (2013). https://doi.org/10.1080/01900692.2012.757619

7. Maerz, S.F.: The electronic face of authoritarianism: E-government as a tool for gaining legitimacy in competitive and non-competitive regimes. Gov. Inf. Q. **33**(4), 727–735 (2016)
8. Stier, S.: Political determinants of e-government performance revisited: comparing democracies and autocracies. Gov. Inf. Q. **32**(3), 270–278 (2015)
9. Kabanov, Y.: Refining the UN E-participation Index: introducing the deliberative assessment using the Varieties of Democracy data. Gov. Inf. Q. **39**(1), 101656 (2022)
10. Taubman, G.: Keeping out the Internet? Non-democratic legitimacy and access to the web. First Monday **7**(9) (2002)
11. Karlsson, M., Adenskog, M.: The case for a broader approach to e-participation research: hybridity, isolation and system orientation. In: Edelmann, N., Danneels, L., Novak, A.-S., Panagiotopoulos, P., Susha, I. (eds.) ePart 2023, pp. 3–14. Springer, Cham (2023). https://doi.org/10.1007/978-3-031-41617-0_1
12. Kersting, N.: Online participation: from 'invited' to 'invented' spaces. Int. J. Electron. Gov. **6**(4), 270–280 (2013)
13. United Nations Department of Economic and Social Affairs United Nations E-Government Survey 2022. United Nations (2022)
14. Lidén, G.: Technology and democracy: validity in measurements of e-democracy. Democratisation **22**(4), 698–713 (2015)
15. Sundberg, L.: Shaping up e-participation evaluation: a multi-criteria analysis. In: Edelmann, N., Parycek, P., Misuraca, G., Panagiotopoulos, P., Charalabidis, Y., Virkar, S. (eds.) ePart 2018. LNCS, vol. 11021, pp. 3–12. Springer, Cham (2018). https://doi.org/10.1007/978-3-319-98578-7_1
16. Pirannejad, A., Janssen, M., Rezaei, J.: Towards a balanced E-participation index: integrating government and society perspectives. Gov. Inf. Q. **36**(4), 101404 (2019)
17. Hofmann, G., et al.: Assessing e-participation indices: a call for more valid measurement. In: Proceedings of the 16th International Conference on Theory and Practice of Electronic Governance, pp. 270–277 (2023)
18. Serdült, U., Hofmann, G., Vayenas, C.: Introducing the DigiPart-index: mapping and explaining digital political participation on the subnational level in Switzerland. In: Proceedings of the 15th International Conference on Theory and Practice of Electronic Governance, pp. 229–236 (2022)
19. North, D.C.: Institutions, Institutional Change and Economic Performance. Cambridge University Press (1990)
20. Milner, H.V.: The digital divide: the role of political institutions in technology diffusion. Comp. Pol. Stud. **39**(2), 176–199 (2006)
21. Bussell, J.: Explaining cross-national variation in government adoption of new technologies. Int. Stud. Quart. **55**(1), 267–280 (2011)
22. Brancati, D.: Democratic authoritarianism: origins and effects. Annu. Rev. Polit. Sci. **17**(1), 313–326 (2014). https://doi.org/10.1146/annurev-polisci-052013-115248
23. Ekman, J.: Political participation and regime stability: a framework for analysing hybrid regimes. Int. Polit. Sci. Rev. **30**(1), 7–31 (2009)
24. Katz, J., Halpern, D.: Political and developmental correlates of social media participation in government: a global survey of national leadership websites. Int. J. Public Adm. **36**(1), 1–15 (2013)
25. Woo, S.Y., Kübler, D.: Taking Stock of Democratic Innovations and Their Emergence in (unlikely) Authoritarian Contexts. Politische Vierteljahresschrift **61**(2), 335–355 (2020). https://doi.org/10.1007/s11615-020-00236-4
26. Norris, P.: Digital Divide: Civic Engagement, Information Poverty, and the Internet Worldwide. Cambridge University Press (2001)
27. Kneuer, M., Harnisch, S.: Diffusion of e-government and e-participation in democracies and autocracies. Global Pol. **7**(4), 548–556 (2016)

28. Geddes, B., Wright, J., Frantz, E.: Autocratic breakdown and regime transitions: a new data set. Perspect. Polit. **12**(2), 313–331 (2014)
29. Kabanov, Y.: The interaction between ICT and authoritarian legitimation strategies: an empirical inquiry. In: Chugunov, A., Khodachek, I., Misnikov, Y., Trutnev, D. (eds.) EGOSE 2020. CCIS, vol. 1349, pp. 184–194. Springer, Cham (2020). https://doi.org/10.1007/978-3-030-67238-6_13
30. Reich, G.: Categorizing political regimes: new data for old problems. Democratization **9**(4), 1–24 (2002)
31. Cho, B., Rethemeyer, R.K.: Whom do we learn from? The impact of global networks and political regime types on e-government development. Int. Public Manag. J. **26**(4), 507–527 (2023)
32. Choi, C., Jee, S.H.: Differential effects of information and communication technology on (de-)mocratization of authoritarian regimes. Int. Stud. Quart. **65**(4), 1163–1175 (2021)
33. Boix, C., Miller, M., Rosato, S.: A complete data set of political regimes, 1800–2007. Comp. Pol. Stud. **46**(12), 1523–1554 (2013)
34. Marshall, M.G., Gurr, T.R.: Polity5: Political regime characteristics and transitions, 1800–2018. Center for Systemic Peace, vol. 2 (2020)
35. Alvarez, M., Cheibub, J.A., Limongi, F., Przeworski, A.: Classifying political regimes. Stud. Comp. Int. Dev. **31**, 3–36 (1996)
36. Bjørnskov, C., Rode, M.: Regime types and regime change: a new dataset on democracy, coups, and political institutions. Rev. Int. Organ. **15**, 531–551 (2020)
37. Schneider, C.: Issues in Measuring Political Regimes. Central European University DISC Working Paper, 2010/12 (2010)
38. Wright, J.: The latent characteristics that structure autocratic rule. Polit. Sci. Res. Methods **9**(1), 1–19 (2019)
39. Munck, G.L., Verkuilen, J.: Conceptualizing and measuring democracy: evaluating alternative indices. Comp. Pol. Stud. **35**(1), 5–34 (2002). https://doi.org/10.1177/0010414002035001001
40. Wigell, M.: Mapping "hybrid regimes": regime types and concepts in comparative politics. Democratization **15**(2), 230–250 (2008). https://doi.org/10.1080/13510340701846319
41. Diamond, L.: Democracy's third wave today. Democracy's Next Wave **110**(739), 299–307 (2011)
42. Huntington, S.P.: Democracy's third wave. J. Democr. **2**(2), 12–34 (1991). https://doi.org/10.1353/jod.1991.0016
43. Lührmann, A., Lindberg, S.I.: A third wave of autocratization is here: what is new about it? Democratisation **26**(7), 1095–1113 (2019)
44. Purcell, S.K., et al.: Authoritarianism. Comp. Polit. **5**(2), 301 (1973). https://doi.org/10.2307/421246
45. Frye, T.: Building States and Markets After Communism: The Perils of Polarised Democracy. Cambridge University Press (2010)
46. Geddes, B., Wright, J., Frantz, E.: How Dictatorships Work: Power, Personalisation, and Collapse. Cambridge University Press, Cambridge (2018)
47. Byman, D., Lind, J.: Pyongyang's survival strategy: tools of authoritarian control in North Korea. Int. Secur. **35**(1), 44–74 (2010)
48. Katchanovski, I., La Porte, T.: Cyberdemocracy or Potemkin e-villages? Electronic governments in OECD and post-communist countries. Int. J. Public Adm. **28**(7–8), 665–681 (2005)
49. Kang, X.: Moving toward neo-totalitarianism: a political-sociological analysis of the evolution of administrative absorption of society in China. Nonprofit Policy Forum **9**(1), 1–8 (2018)

50. Wehrey, F.: The Authoritarian Resurgence: Saudi Arabia's Anxious Autocrats (2015). https://carnegieendowment.org. Accessed 13 June 2024
51. Lee-Geiller, S.: Technology married to good governance and diversity: explaining e-participation preparedness in government. Technol. Forecast. Soc. Chang. **201**, 123218 (2024)

Sustainable eParticipation Through Lightweight Democracy?

Marius Rohde Johannessen[1(✉)], Noella Edelmann[2], and Lasse Berntzen[1]

[1] School of Business, University of South-Eastern Norway, Borre, Norway
mj@usn.no
[2] University of Continuing Education Krems, Krems an der Donau, Austria

Abstract. Sustainable eParticipation is defined as participation projects that last and become part of the democratic process. Many eParticipation projects fail in this regard. The eParticipation literature presents many studies and test, but few report on the aspects that can make tools and initiatives sustainable over time. The past decade's rise of populist rhetoric and populist parties in the west shows the urgency of finding good answers to the question of sustainability in eParticipation. In this paper, we argue that lightweight eParticipation initiatives, with low demands on time and effort from citizens, can contribute to sustainable eParticipation. We test this assumption through an evaluation of a Norwegian eParticipation system for low-threshold rapid feedback. Our findings indicate that while the system showed promise and was well received, there were still issues with sustainability, especially related to strategy and financing.

Keywords: eParticipation · implementation · survey · pilot study · sustainable participation · evaluation

1 Introduction

The state of democracy has been a concern for decades. The current political communication landscape is characterised by populism and anti-elitist sentiment [1], a public divided on topics such as Trump vs Sanders in the US, Brexit in the UK, immigration, climate change and the environment following a logic of "us vs them", with little room for reasoned debate in a Habermasian sense [2]. In Europe we observe what Minkenberg terms the growth of the "radical right" (a somewhat politicized term) in countries such as Italy or Austria, a general shift to the right following the European elections in 2024, whilst some countries, for example, Hungary or Poland are changing the legal system, introducing "anti-gay zones" and attacking the media [3]. The past two decades have promised and demonstrated that the Internet and social media can be used for politics: people are political and a plethora of social movements and organizations are mobilised and supported though the us internet-based tools [4].

Research and implementation of new forms of public participation often have a techno-optimistic vision, where digital technology fosters and supports democracy, accompanied by strategies for the adoption, implementation, and institutionalization of eParticipation initiatives, taking into consideration the spectrum of participation possibilities, stakeholders and their diverse roles, as well as citizens' degrees of power [5]. Between complex, tailored systems and the anarchy of social media, Hibbing and Theiss-Morse [6] claim that most people want to be heard but are not interested in taking the time to really understand and read up on complex issues. They argue that a good approach to participation is to avoid information overload by presenting information spanning hundreds of pages, and instead, ask simple questions about issues citizens can form an opinion on. Yet there is significant evidence that shows that populism and extreme opinions are more prevalent here and have been quicker at adopting digital channels and adapting their communication to the characteristics of the medium [7], as seen, for example with the Italian's Moviemento 5 Stelle's use of social media. There is a need for research on the sustainability of eParticipation processes and outcomes, that is, to be able to attract and retain users over time d ensure long-term sustainable outcomes. Wirtz and Daiser [8] believe that one reason why the sustainability and success of eParticipation activities is limited is because strategic approaches are based on a diffuse definition of eParticipation. To develop an eParticpation strategy, they suggest defining eParticipation as "a participatory process that is enabled by modern information and communication technologies, includes stakeholders in the public decision-making processes through active information exchange, and thus fosters fair and representative policymaking" (p.3). Based on this definition, their Integrated Strategic E-Participation Framework considers the eParticipation targets, the forms, types of strategies, instruments, and the demand groups to address the key dimensions accountability, transparency, technology and stakeholders. This framework has been successfully applied for example, in the development of new national participation guidelines in Austria [9], but we suggest the need to consider sustainability in terms of e.g. [10] who describes sustainability as "the ability (…) to uphold or support, i.e. sustain something considered valuable", a definition that allows an understanding of sustainability as the capacity to maintain an entity, outcome or process over time, without producing irreversible, adverse effects on the environment upon which it depends. In this paper, we present the findings from an evaluation of a Norwegian eParticipation system, built on this philosophy, asking simple questions, and providing politicians with a temperature gauge of public opinion. The research question we ask is therefore: *Can light-weight democracy contribute to sustainable eParticipation?*

2 Related Research

In the 1990s open discussion forums were popular but had limited success [11]. Later, complex and advanced participatory systems, designed top-down for decision-makers to receive input on concrete issues, were developed (see e.g., [12, 13]). The EU's 7th Framework Programme for Research funding made calls for the development of eParticipation, and in the early to mid-2010's many different tools and evaluations of pilot projects [14] were presented in academic journals and at conferences (e.g., [12, 13]). The evaluations concluded that these technical systems provide excellent feedback, but are complex and time-consuming, and do not attract enough participants and comments were considered to have "failed" [14]. Another phenomenon associated with eParticipation is the "Matthew Effect", that is, that that those who already have status can gain more, whereas those without status struggle more to gain it [15]. The challenges and failures associated with implementing eParticipation are usually attributed to social, administrative, and institutional factors rather than technical ones [16]. Even the inclusion of game elements or social media, that are supposed to lower the threshold for participation and encourage people to discuss politics [17, 18], neither contributes to increasing the quality of communication nor helps to manage the conversation and extract meaningful information [2, 19]. At the same time, social media polarization, fake news, bots spreading propaganda, and increasing number of activist web sites position themselves as alternatives to mainstream media [20, 21]. The so-called alt-right find each other in online for a such as 4chan, Tumblr and 8chan, and coordinate campaigns against political opponents and disseminate disinformation [22]. The Norwegian paper *Morgenbladet*[1], in collaboration with the breaking bad research project, V-dem and Freedom House, shows that several countries, including Western European ones, are moving away from liberal ideals [23], as seen in responses such as the "Yellow Vests" in France, the negative attitudes towards migration during the 2015 refugee crisis in Europe, the development of an "illiberal democracy" in Eastern Europe, or the election of populists as president or prime minister.

Previous research has already shown that eParticipation systems need a clear purpose and form [24], concrete outcomes [25] and feedback mechanisms so that citizens see the impact and outcome of their participation [26].

2.1 Models of Democracy and Participation

Democracies have an obligation to involve citizens through elections, political parties [27, 28] and citizen/politician dialogue in various channels and media within the frames of representative democracy [29]. There are several models of democracy, with different normative criteria for participation. Ferree and colleagues [30], for example, describe four different models of democracy: representative liberal, participatory liberal, discursive, and constructionist. The models outline the amount of citizen participation, based on "who should speak, the content of the process (what), style of speech preferred (how), and the relationship between discourse and decision-making (outcomes) that is sought (or feared)" (p. 290). While some countries focus only on voting, others, such as Norway or Austria, see it as a democratic value that citizens engage in dialogue and are involved in decision-making between elections (participatory liberal model), and participation in the public debate is seen as a value in and of itself (Habermasian discursive model). The Norwegian constitution (§100, part 6) states that "government is required to facilitate open and rational public discourse". Even so, membership in political parties is in decline, with only 7% of the adult population being member of a political party [31]. Clearly, new ways of encouraging citizen engagement are needed.

Citizen involvement has a number of positive effects on democracy: it increases issue knowledge, civic skills, public engagement, and it contributes to the support for

---

[1] https://www.morgenbladet.no/aktuelt/demokrati/2019/03/22/lovendringer-I-europa-som-svekker-demokratiet/.

decisions. The widely known OECD's (2001) framework on e-Participation is includes i) information, ii) consultation and iii) active participation. Macintosh (2004) adds three levels of e-Participation: (1) e-enabling, (2) e-engaging and (3) e-empowering. Participation can also be extended into four stages: information, consultation, co-operation and co-determination [32]. Each stage also includes a specific range of activities with varying outcomes and effects, targeting different democratic ideal-types [33]. First, from the perspective of civil society, participation is understood as beneficial, as the public can express their ideas, interests and needs in regard to decision-making processes, be active and access comprehensive information and different points of view; second, from a political-administrative perspective, participation procedures can help address problems and complaints, lead to the development of viable solutions, and increase trust in public administrations. Third, from a political perspective, participation can help provide an overview of the needs different population groups have and are they may be help and supported through communication and dialogue with them. This final perspective may support citizens' interest in politics and motivate them to participate. Michels [34] argues that citizen participation: "gives citizens a say in decision-making (influence); contributes to the inclusion of individual citizens in the policy process (inclusion); encourages civic skills and virtues (skills and virtues); leads to rational decisions based on public reasoning (deliberation); increases the legitimacy of decisions (legitimacy) (p. 279). Whilst it has often been stated that enabling meaningful participation must consider the stakeholders who are affected or interested, the nature of the topic, the need for cooperation between the citizens, the public sector and politicians [35], it also means that researchers should be clear about the type of democracy the eParticipation activity supports.

The eParticipation activity presented in this paper is designed as a tool for consultation, that is where "the involved parties (citizens, companies, NPOs) [...] express their opinion on questions posed, make proposals or official statements on submitted drafts. [...] communication flows mutually between the public and its representatives in legislation (MPs) and/or the stakeholders in public administration" [32] The extent of civil society influence on the decision can differ considerably, depending on the use of the tool, such as simple ePolling or more extensive eDeliberation. A consultation should be implemented at an early stage, where politicians ask questions and citizens give answers. The expected outcome is increased civic engagement and general democratic effects, depending on how the politicians decide to use the feedback from the activity.

2.2 Lightweight Democracy and Nuanced Forms of Participation

In their book *Stealth Democracy,* Hibbing and Theiss-Morse [6] point out that many citizens are tired of politics and the political debates found in the media. Citizens reported being tired of the conflict, constant debates and difficult-to-understand political compromise. They have little interest in how democracy works in practice, and do not wish to become too involved. Rather, citizens want to be able to express their opinion and be heard without having to spend time reading long policy documents or become too involved. The contribution of participation to democracy differs according to type of democratic tool or innovations, so deliberative forums and surveys appear to be better at promoting the exchange of arguments, whereas referendums and participatory policy making projects are better at giving citizens influence on policy making and involving more people [34]. FixMyStreet - type services that have individual democratic benefits as they are related to their own areas of interest and life are more popular than long, procedural participation tools [36] that may have benefits to democracy as a whole.

Applying a pragmatic view on participation, we argue that eParticipation activities such as the one presented in this paper have the potential to be the missing link between the open and unstructured debate found in social media and the more tailored and complex systems for eParticipation. Lightweight, "stealth" participation in the form of surveys, data analysis through sensors or apps such as FixMyStreet allow citizens to participate in a way that gives valuable insights to decision-makers, without having to spend too much time and effort. Over the past few years, several studies have examined how lightweight participation can contribute to democracy There are alternative forms to participation, such as lurking which can have extensive impact on eParticipation [37]. Activities such as online lurking may contribute to sustaining valuable e-participation, as even online participation that is not visible may have an impact through activities such as listening, acting as an audience, using, propagating, and sharing knowledge [37]. Lutz and Hoffmann [38], argue that there are many nuanced dimensions of "non-, passive and negative participation" (p. 1), and participants may engage in both willingly and unwillingly ("agency") and with both intentional and unintentional positive and negative effects ("valence"). Amna and Ekman argue that the need to consider political passivity as in three distinctive forms, the "standby citizen", the "unengaged" and the "disillusioned" citizens. Each group of citizens display different political behavior, which does not only imply the need for a new analytical framework to study this phenomenon, but also has implication for eParticipation research, implementation and evaluation [34, 39].

2.3 From Strategic to Sustainable eParticipation

Wirtz et al. [8] state that eParticipation processes should be systematically addressed by eParticipation providers. They need to consider the strategic and organizational factors and the environmental drivers, that is, both the internal and the external factors that may have an impact on eParticipation. This means that the selection and design of eParticipation must consider the stakeholders' preferences, especially when considering technology-based interaction. As one of the key elements is the participation provider's ability to choose the right channels, as these are the interface to the public stakeholders, participation can be based on 3 strategies: isolated, combined, and integrated. These strategies differ in terms of the integration and coordination of the instruments, the organizational factors, the targets/objectives to be achieved and forms of participation that can range from informing to empowering participants.

The European Commission [40] points out that digital policies should empower people and businesses, support the transition to a sustainable economy, focus on digital literacy, data infrastructure and data-processing technologies, so eParticipation is key to having a sustainable impact on democracy. Sustainability is tightly interlinked with the digital transformation in Europe [41] and advances in ICTs have opened opportunities to transform the relationship between public administrations, citizens, and other stakeholders, but requires the "involvement and commitment of the public and of all stakeholders" [40]. There has been research on sustainable eParticpation [42], but it is limited and often the literature focuses on sustainable development instead [41]. Whilst sustainable development can be understood as the "promotion of societal transformation processes by governments, market actors and civil society" [43], in the public sector, sustainability is often "understood as the effort needed to maintain an object or an action over a prolonged period" [44]. Sustainability analysis usually focuses on six dimensions: stakeholders' ownership, institutional compliance: financial autonomy, socio-cultural integration: technical feasibility and continuity over time [42]. We therefore argue that eParticipation strategies and implementation must consider sustainability in order to achieve the set outcomes over time, where the outcome is considered valuable and can be achieved without producing irreversible, adverse effects on the environment as argued by Türke [10].

3 Research Approach

The objective of this paper is to examine the sustainability of eParticipation through the evaluation of a lightweight participation initiative. To answer the research question, we applied a case study mixed-methods approach grounded in interpretivism.

Casing: Since 1993, the Norwegian "power and democracy" project has been concerned with falling membership in political parties, falling interest in broad movements in favour of single-issue politics, and a fall in voter turnout [45]. In Norway, most citizens still report high levels of trust in the political system and institutions, but a significant minority is less trusting and choosing not to vote in elections [46]. In addition, in recent years, several developments such as fake news, bots, polarization, right- (and left-) wing activism, echo chambers and a plethora of new online news sources with an agenda are on the rise [20, 21, 47]. The objective of the company [anonymous for submission] described in this case study was to fight against these trends, and to provide a voice to people who would otherwise not be heard (interview with company founders).

Data Collection: The initial data collection period lasted approximately 12 months in 2018–19, including development, implementation, and pilot testing. For the development and implementation phases, data collected are mainly qualitative, gained in the form of participant observation [48] in project meetings and workshops with municipal, volunteer - and private sector stakeholders. 12 mayors and 50 politicians were present in these workshops, as well as representatives from non-governmental organizations and the media. E-mail interviews were conducted with politicians in the pilot municipalities, as well as informal talks with researchers focussing on media, democracy, and digitalization. In this process, two of the authors had an active role in shaping the eParticipation activity. Academic staff from the University of South-Eastern Norway validated a survey and tested it on five random users prior to distribution. Development of the app was an iterative process during this period. Input from data collection informed the app developers, and versions of the app (from wireframes to early prototypes) were presented to informants during workshops and project meetings.

The app was tested in five Norwegian municipalities, with a limited number of citizens. An evaluation survey was sent to the pilot participants ($N = 389$) after the app was tested, and 189 valid answers were returned. The participants were citizens and high school students who had volunteered to test the initiative.

In 2020, a second, and larger, research innovation project was funded by the Norwegian Research Fund. The objective of this project was to further develop the original eParticipation initiative and have it ready for implementation by 2022. For the case study presented here, additional observational data from meetings with the company, research team and meetings with municipalities implementing the system is included. This is presented in Sect. 4.4.

Data Analysis: The analysis of the data was based on the ideas of lightweight democracy and the company's idea of reaching those people who do not normally participate in political processes. Field notes and interview questions were structured and coded accordingly. The survey was based on questions developed from the acceptance of technology literature and inspired by the constructs in the Unified Theory of Acceptance and Use of Technology (UTAUT) [49], adding trust and demographic variables as these have shown to be relevant for technology acceptance [50]. Wirtz et al.'s [8] Integrated strategic eParticipation framework was used to uncover how lightweight democracy may contribute to sustainable participation and is presented in Sect. 4.4.

4 Findings

4.1 Presentation of the Case

In January 2018 a small start-up company wanted to discuss the possibilities of a system for lightweight democracy. The start-up, here anonymized as "Democracy inc.", consists of people with a broad background from business, the voluntary sector (youth sports) and the media industry. They used a local UX/web company to build the app and ecosystem and focused their own efforts on their wide network of possible partners, where they were extremely active in building a network of business, government, NGO's, and political partners who stated their support for this type of initiative. This network played an important role in the process from idea to realization, as it opened doors for the company and made it easier to recruit participants. The start-up company had been hired by their local municipality to organize a "state your concerns on a post-IT note" session, and were amazed by the strong reactions from citizens, such as "finally someone wants to listen to us". This was the starting point for thinking about a digital activity for eParticipation.

The company's objectives for the system were to 1) Create an activity that would ensure the "silent majority" could participate in political processes in an easy way. The silent majority was defined as those who are not represented via non-profit/non-government organizations (who by law needs to be invited to policy hearings) and rarely raise their hand in public meetings. Youth was targeted as being especially important. 2) Develop an app where the municipality can consult with citizens on current issues. 3) Citizens/participants only need to spend two-three minutes of the time, so easy to use that you can do it in the checkout queue at the supermarket.

The participation system architecture is quite simple, with a common database, a back-end system for generating questions and analyzing responses, and a mobile app dialogue tool. Upon opening the app, users select their municipality. A brief video from the mayor then welcomes them, explaining the app's purpose and usage. Users have the option to sign up for all surveys or only those pertaining to their interests. Participation involves answering a straightforward survey consisting of 5–10 questions, with an option for more detailed responses at the end. After completing the survey, users are shown a thank-you screen and provided with contact details should they wish to discuss additional issues or provide further comments. Users also have the ability to view responses from other participants. This is an important aspect of the app, as seeing others' opinions can aid mutual understanding [51]. After completing a round of questions, participants receive feedback on how the results are being used. This includes ideas on how the input can be applied by the municipality. Figure 1 presents screenshots from the app.

Fig. 1. Screenshot from the app showing questions (1, 2), thank you note from the mayor (3) and a summary of responses (4).

4.2 Workshops – Feedback from Major Stakeholders

The workshops were conducted iteratively with the development of the app and uncovered a desire among attendees to engage in broader discussions about participation and democracy. Many participants expressed that they found the most valuable aspect of the workshop to be the opportunity it provided for them to delve into normative and fundamental democratic issues, topics that are frequently overlooked in the hustle and bustle of daily politics. In addition to these points, participants brought up the following concerns:

The Need for Improved Communication: Most participants concurred on the need to engage a wider demographic, particularly younger individuals, and non-members of traditionally consulted organizations. While older citizens often interact with local politicians or participate in public meetings, younger ones tend not to. Attempts to use social media have been hindered by an unsuitable tone and style for civil, rational debate. There was agreement on the need for more efforts to connect with the young and the silent, with initiatives like this receiving political approval.

Ownership and Organization: Much of the discussion was related to practical issues such as ownership, financing, and organization. Specifically, the following issues were raised: *Who should be invited to participate?* Should there be a representative panel, or should everyone be invited to answer every survey? Ensuring representative and valid answers was also an issue. *Who should be in control of the system and act as the figurehead?* This raised heavy debate on the pros and cons of having the mayor act as front and apparently controlling the questions being asked. This was discussed during several workshops, and the consensus seemed to be that mayors should front the app, as a human face is better than a municipal logo. A committee of politicians from the parties represented in the municipal government should be responsible for the back-end, so that party politics do not influence the questions and topics being addressed via the app. An illustration of how controversial these issues can be is that in one of the municipalities, a member of the political opposition wrote an angry and critical letter to the local newspaper about how the position "used the app to validate their own policies" and what he saw as a lack of transparency in who could participate and how answers were being used.

Formulation of Questions: Creating unbiased and relevant questions for a survey is difficult, and one of the participants' major areas of concern. The consensus was that simple questions about what [do you think about...] and how [should we handle...] should be used to ask questions about issues being discussed in the municipal government. This represents a top-down approach rather than an open discussion but is about topics high on the agenda in local news and local social media groups. Ideological issues should be avoided, as simple survey questions are not suitable for discussion.

Criticism: While most workshop participants were positive, some NGOs raised concerns about the democratic outcomes of Democracy inc.'s system, in particular, the kind of democracy this would facilitate. They discussed what it meant to "be heard" vs having an actual effect on policy outcomes, if the app was really any different from a regular survey and if this type of eParticipation activity involves those who are not participating and the younger citizens or rather is a form of tokenism. The results from the workshops as well as previous research would suggest that this is "consultation" democracy. As described above, Democracy inc. is not designed to be a deliberative system. The app does not aim to be polling station, rather, it allows for a quick and easy temperature check on current issues. The expected outcome of participation and public interest is discussed further in the next section.

4.3 Survey – Feedback from Citizen Participants

After the app was tested in the first pilot study (five municipalities, with a limited number of self-selected participants), we distributed an evaluation survey to participants, asking them about their experience with using the app, as well as their willingness to continue using it. Here, we present a summary of the survey findings:

Background and Demographic Variables: The respondents comprised 47% females and 53% males. Age distribution was as follows: 35% aged 15–24, 56% aged 25–39, 56% aged 40–54, and 36% aged 55 or older. Age had minimal impact on responses. Younger respondents were more active on social media and receptive to gamification, while older respondents were more likely to sign petitions. Young adults (25–39) were least politically active. High school students were most positive towards the app, frequently contacting the mayor and Democracy Inc. The sample investigated has a somewhat higher level of education compared to the total Norwegian population[2], especially at master's level (sample: 29%, population in 2017: 10%). The respondents are also somewhat more politically active than the general population. 5% of the general population have written a political letter to the newspaper, vs 13% of the respondents. 6% have attended a demonstration, vs 8% of the respondents. 12% of the general population and 25% of our respondents say they have contacted a politician directly.

Attitudes to Personalisation: An important element that emerged from the workshop discussions was using the mayor as the front-figure of the app, vs using the municipality logo. This was heavily debated, with politicians having strong opinions for and against. For the pilot, we chose to have the mayor present the app, as there is a general trend towards politics becoming more person-oriented [52]. 70% of the respondents saw it as positive that the mayor's face was the first thing to greet them in the app. 65% also respond positively to the statement "the mayor is a unifying force in my municipality". In Norway, mayors are expected to be mayor first and party member second, so our survey confirms that this is true for the five pilot municipalities.

Expectations Towards Outcome: As previous research has shown [25], Clarity about participation outcomes is crucial. 65% of the participants reported that they expect the app outcome as "being heard and having their opinions taken into account as part of the formal hearing process in policy-making", with only 6% expecting their input to have a direct consequence. 29% had no expectations or were unsure what to expect.

Trust and Intention to Use: Trust, or a lack of trust, is one possible explanation for the current wave of populism and anti-elitist sentiment [46]. The show high levels of trust in local politicians and information from local government. Trust is positively correlated with participants' intention to continue using the system. When asked if they intend to continue using the app if it becomes available after the pilot, 70% somewhat or fully agree that they would like this, while 25% are neutral/unsure.

4.4 Achieving Sustainable eParticipation

In 2020 the project was extended and received further funding for development and implementation. The purpose of this second project was to implement the initial pilot into a verified and research-based concept consisting of the app-based system, a methodology for use and a framework for implementation. The system was to be implemented and tested on a larger scale, in schools and municipalities. A major new feature was the use of a synchronous/same place and time module for the initiative, like Kahoot and Mentimeter. However, the pandemic lockdown disrupted the process, and the eParticipation activity was only tested in three schools and one municipality. In addition, a key member of the company passed away unexpectedly. This led to the project being cancelled and

---
[2] General population statistics from Statistics Norway, www.ssb.no.

the company going bankrupt. In other words, the project failed to become sustainable. Why did this happen? The CoVID-19 pandemic played a major part, as workshops and physical meetings had to be cancelled. But there were also other factors, which can be related to Wirtz et al.'s [8] framework:

eParticipation Targets: The objective for the project was to give a voice to the silent majority, but beyond this it quickly became clear that objectives for using the system varied between participants. Participant involvement is essential for sustainability, as shown in [44].

eParticipation Forms: Form refers to the level of participation being targeted (information, consultation, involvement, collaboration, empowerment). There were issues here related to the lack of agreement on what the targets should be. Especially in the municipality, there was no clear agreement on the form of eParticipation. Some understood it as a consultation, others as a survey tool for measuring public sentiment. The municipality did not communicate well to participants what the input would be used for. Part of the problem for targets and forms was the fact that the project and municipalities involved failed to agree on a clear definition of democracy and on the outcomes to be achieved. In the municipality that implemented the system, participants and politicians had different views on what the activity was supposed to achieve.

eParticipation Strategies: A major problem was the disruption caused by the CoVID-19 pandemic. The researchers were supposed to examine various strategies for implementation and outcomes, but this was made impossible because of the CoVID-19 pandemic. The different stakeholders involved in the second project (researchers, company, municipalities, and schools) had varying and to some extent competing interests, and the full-scale municipality implementation was not planned with research in mind. As the CoVID-19 pandemic placed restrictions on meetings, it was difficult to properly resolve these issues. Further, the methodology for use and follow-up of data collection was never fully developed. Long-term financing is an important aspect of sustainability [44], and indeed it is the lack of ownership and financing that led Democracy inc. to bankruptcy. The app was well-received and liked, but although Democracy Inc. tried several financing models, such as licenses, selling the system, and transferring ownership to KS, an organization handling software for municipalities, there was no willingness to pay, or provide funding mechanisms to secure long-term sustainability.

5 Conclusion and Future Work

The past decade's rise of populist rhetoric and populist parties in the West shows the urgency of finding good answers to the question of sustainability in eParticipation. Therefore, in this paper we examined sustainability in eParticpation, with a particular focus on if and how lightweight approaches to participation are sustainable.

The literature on eParticipation indicates that lightweight initiatives that do not place a lot of work on participants so they can attract more participants and be easier to sustain over time. This study presents the analysis of a Norwegian system for lightweight participation. The findings from the first phase of pilot testing (2018–2019) indicate that the test participants were mostly happy with the system, had moderate expectations about how their input should be used (in line with representative democratic ideals), and a majority will use the system if it becomes available. The findings described in Sect. 4.2 illustrates the complexity of eParticipation, even for lightweight initiatives. Expectations, outcome, organization, use, and financing are important factors that contribute to the sustainability of eParticipation and must be clearly defined and agreed upon within each municipality.

There were many discussions on issues related to targets, forms, and strategies [8]. In the workshops and meetings, the need to engage more people from varied backgrounds is something most, if not all, participants agreed on, so it seems "targets" is not the major issue – at least not until we discuss specific cases. Then, questions were raised about who should be invited to participate (open for all, a carefully selected panel or something in between?) and the implications of various recruitment strategies. There were also discussions on who should be in control of the system and the responsibility for raising issues to be answered through the app, and on how to create unbiased and relevant questions for discussion, as well as on how to use the results, level of participation and type of democracy supported. The importance of clarifying these issues were made clear when a member of the opposition in one of the municipalities wrote an angry letter to the local newspaper, which shows how important it is to be transparent on targets, forms, and strategies if we want initiatives to be successful and sustainable, but our findings also indicate that this, will it seems simple, is a long and complicated discussion in a democratic forum such as a municipal council.

Section 4.4 describes the second phase of the project, where the objective was related to implementation and sustainability. Despite promising findings from the pilot in terms of interest and acceptance, lightweight tools must also consider the targets, forms, and strategies in close cooperation with stakeholders. There were challenges related to this, that we unfortunately were not able to address as the project was cancelled. Here too, the results show that financial sustainability remains a challenge for eParticipation. Even when most stakeholders want to use the system, it is difficult to establish a lasting source of income as municipalities do not have funding for democracy tools.

We conclude that the ideas of lightweight participation can be valuable contributions to furthering local democracy, but not necessarily the only form of participation. Lightweight participation can help attract more citizens to participate, and perhaps also help with raising political engagement and recruit people to more "heavy" participation initiatives. Further, efforts related to strategy and financing need to consider the different aspects and require close attention and coordination. As the findings presented here are from a single case in one country, we call for more research into the sustainability of lightweight participation tools in other contexts. For example, it would be especially interesting to see studies of similar initiatives in low and middle-income countries. Finally, we realize that lightweight tools have limited potential for the more involved forms of participation (collaboration and empowerment). Hence, further research is needed to identify how to balance complexity, citizens' willingness to spend time and effort as well as the democratic values and levels of participation. This could help identify a system of eParticipation tools that can attract citizens to lightweight participation and perhaps motivate some of them to move on to more time-consuming and involved forms of participation. Identifying such a system and how to finance it, could potentially lead to lasting and sustainable eParticipation that can act as a barrier against current populist trends.

Acknowledgements. This study was funded by the Research council of Norway, project 310122. The submission is an extension of a reflections paper from ePart 2020.

References

1. Schroeder: Social Theory after the Internet: Media, Technology and Globalization. UCL Press (2018)
2. Johannessen, M.R.: Genres of participation in social networking systems: a study of the 2017 Norwegian parliamentary election. In: Edelmann, N., Parycek, P., Misuraca, G., Panagiotopoulos, P., Charalabidis, Y., Virkar, S. (eds.) Electronic Participation. LNCS, pp. 64–75. Springer, Cham (2018). https://doi.org/10.1007/978-3-319-98578-7_6
3. Minkenberg, M.: The Radical Right in Eastern Europe: Democracy Under Siege? Springer, Heidelberg (2017)
4. Markoff, J.: Waves of Democracy: Social Movements and Political Change, 2nd edn. (2015). https://doi.org/10.4324/9781315631202
5. Steinbach, M., Sieweke, J., Süß, S.: The diffusion of e-participation in public administrations: a systematic literature review. J. Organ. Comput. Electron. Commer. **29**, 61–95 (2019). https://doi.org/10.1080/10919392.2019.1552749
6. Hibbing, J.R., Theiss-Morse, E.: Stealth Democracy: Americans' Beliefs About How Government Should Work. Cambridge University Press, Cambridge (2002)
7. Zhuravskaya, E., Petrova, M., Enikolopov, R.: Political effects of the Internet and social media. Ann. Rev. Econ. **12**, 415–438 (2020). https://doi.org/10.1146/annurev-economics-081919-050239
8. Wirtz, B.W., Daiser, P., Binkowska, B.: E-participation: a strategic framework. Int. J. Public Adm. **41**, 1–12 (2018). https://doi.org/10.1080/01900692.2016.1242620
9. Edelmann, N., Albrecht, V., Parycek, P.: Strategic Public Participation in the Digital Age: The Case of the Austrian "Green Book." Edward Elgar Publishing (2024)
10. Türke, R.E.: Sustainable governance. In: Systemic Management for Intelligent Organizations: Concepts, Models-Based Approaches and Applications, 9783642292446, pp. 237–247 (2012). https://doi.org/10.1007/978-3-642-29244-6_14/COVER
11. Sæbø, Ø., Rose, J., Molka-Danielsen, J.: EParticipation: designing and managing political discussion forums. Soc. Sci. Comput. Rev. **28**, 403–426 (2010)
12. Liddo, A., Shum, S.B.: New ways of deliberating online: an empirical comparison of network and threaded interfaces for online discussion. In: Tambouris, E., Macintosh, A., Bannister, F. (eds.) ePart 2014. LNCS, vol. 8654, pp. 90–101. Springer, Heidelberg (2014). https://doi.org/10.1007/978-3-662-44914-1_8
13. Porwol, L., Ojo, A., Breslin, J.: A semantic deliberation model for e-Participation. In: Innovation and the Public Sector (2014). https://doi.org/10.3233/978-1-61499-429-9-40
14. Taudes, A., Leo, H.: Determinants of the willingness to contribute to an econsultation. In: Innovation and the Public Sector (2014). https://doi.org/10.3233/978-1-61499-429-9-13
15. Observing and Negating Matthew Effects in Responsible Research and Innovation Transition I ON-MERRIT I Project I Fact sheet I H2020 I CORDIS I European Commission. https://cordis.europa.eu/project/id/824612. Accessed 14 June 2024

16. Chadwick, A.: The political information cycle in a hybrid news system: the British prime minister and the "Bullygate" affair. Int. J. Press Polit. **16**, 3–29 (2011). https://doi.org/10.1177/1940161210384730
17. Enli, G.S., Skogerbø, E.: Personalized campaigns in party-centred politics: Twitter and Facebook as arenas for political communication. Inf. Commun. Soc. (2013). https://doi.org/10.1080/1369118X.2013.782330
18. Elvestad, E., Johannessen, M.R.: Facebook and local newspapers' effect on local politicians' popularity. Northern Lights: Film Media Stud. Yearb. **15**, 33–50 (2017). https://doi.org/10.1386/nl.15.1.33_1
19. Majumdar, S.R.: The case of public involvement in transportation planning using social media. Case Stud Transp Policy (2017). https://doi.org/10.1016/j.cstp.2016.11.002
20. Sunstein, C.R.: Republic Divided: Divided Democracy in the Age of Social Media. Princeton University Press, Princeton (2018)
21. Allcott, H., Gentzkow, M.: Social media and fake news in the 2016 election. J. Econ. Perspect. **31**, 211–236 (2017)
22. Comisión Europea: Joint Communication to the European Parliament, the European council, the Council, the European Economic and Social Committee and the Committee of the Regions (2019). https://doi.org/10.1007/s13398-014-0173-7.2
23. Reinertsen, M.B., Jakobsen, H.Ø., Belgaux, C.: F-ordet (2019)
24. Hurwitz, R.: Who needs politics? Who needs people? The ironies of democracy in cyberspace. In: Jenkins, H., Thorburn, D. (eds.) Democracy and New Media, pp. 101–112. MIT Press, Cambridge (2003)
25. Kolsaker, A., Kelly, L.L.: Citizens' attitudes towards e-government and e-governance: a UK study. Int. J. Public Sect. Manag. **21**, 723–738 (2008)
26. Kolsaker, A.: Third way e-government: the case for local devolution. In: Böhlen, M., Gamper, J., Polasek, W., Wimmer, M.A. (eds.) TCGOV 2005. LNCS (LNAI), vol. 3416, pp. 70–80. Springer, Heidelberg (2005). https://doi.org/10.1007/978-3-540-32257-3_7
27. Dewey, J.: The Public and its Problem. Swallow Press/Ohio University Press, Athens (1927)
28. Oppenheim, F.E.: Democracy—characteristics included and excluded. Monist **55**, 29–50 (1971). https://doi.org/10.2307/27902204
29. Brooks, C., Manza, J.: Why Welfare States Persist: The Importance of Public Opinion in Democracies. University of Chicago Press, Chicago (2007)
30. Ferree, M.M., Gamson, W.A., Gerhards, J., Rucht, D.: Four models of the public sphere in modern democracies (2002). https://doi.org/10.1023/A:1016284431021
31. SSB: No Title
32. Parycek, P., Edelmann, N.: eParticipation and edemocracy in Austria: projects and tenets for an edemocracy strategy. In: 1st International Conference on eGovernment and eGovernance, Ankara (2009)
33. Sæbø, Ø., Rose, J., Skiftenes Flak, L.: The shape of eparticipation: characterizing an emerging research area. Gov. Inf. Q. **25**, 400–428 (2008)
34. Michels, A.: Innovations in democratic governance: how does citizen participation contribute to a better democracy? Int. Rev. Adm. Sci. **77**(2), 275–293 (2011). https://doi.org/10.1177/0020852311399851
35. Arbter, K., Trattnigg, R., Kallinger, M.: Standards der Öffentlichkeitsbeteiligung, Wien (2009)
36. Berntzen, L., Johannessen, M.R., Böhm, S., Weber, C., Morales, R.: Citizens as sensors: Human sensors as a smart city data source. In: SMART 2018 : The Seventh International Conference on Smart Cities, Systems, Devices and Technologies (2018)
37. Edelmann, N., Krimmer, R., Parycek, P.: How online lurking contributes value to E-participation: a conceptual approach to evaluating the role of lurkers in e-participation. In: 2017 4th International Conference on eDemocracy and eGovernment, ICEDEG 2017, pp. 86–93 (2017). https://doi.org/10.1109/ICEDEG.2017.7962517

38. Lutz, C., Hoffmann, C.P.: The dark side of online participation: exploring non-, passive and negative participation. Inf. Commun. Soc. **20**, 876–897 (2017). https://doi.org/10.1080/1369118X.2017.1293129
39. Edelmann, N.: Online Lurking: Definitions, Implications, and Effects on E-participation (2017). https://www.researchgate.net/publication/318274273_Online_Lurking_Definitions_Implications_and_Effects_on_E-participation
40. European Commission: Decision (EU) 2022/2481 of the European Parliament and of the Council of 14 December 2022 establishing the Digital Decade Policy Programme 2030. European Commission (2021)
41. United Nations: Transforming Our World: The 2030 Agenda for Sustainable Development United Nations United Nations Transforming Our World: The 2030 Agenda For Sustainable Development (2015)
42. Molinari, F.: On sustainable eparticipation. In: Tambouris, E., Macintosh, A., Glassey, O. (eds.) Electronic Participation, pp. 126–139. Springer, Heidelberg (2010). https://doi.org/10.1007/978-3-642-15158-3_11
43. Lange, P., Driessen, P.P.J., Sauer, A., Bornemann, B., Burger, P.: Governing towards sustainability—conceptualizing modes of governance. J. Environ. Plann. Policy Manag. **15**, 403–425 (2013). https://doi.org/10.1080/1523908X.2013.769414
44. Edelmann, N., Virkar, S.: The impact of sustainability on co-creation of digital public services. Adm. Sci. **13**(2), 43 (2023). https://doi.org/10.3390/admsci13020043
45. Østerud, Ø., Engelstad, F., Selle, P.: Makten og demokratiet. En sluttbok fra makt-og demokratiutredningen. Gyldendal akademisk, Oslo (2003)
46. Kleven, Ø.: Nordmenn på tillitstoppen i Europa. Samfunnsspeilet. 2 (2016)
47. Nagle, A.: Kill all normies: Online culture wars from 4chan and Tumblr to Trump and the alt-right. John Hunt Publishing (2017)
48. Becker, H.S., Geer, B.: Participant observation and Interviewing: A Comparison. Hum Organ. 16 (1957)
49. Venkatesh, V., Morris, M.G., Davis, G.B., Davis, F.D.: User acceptance of information technology: Toward a unified view. MIS Q. 425–478 (2003)
50. Naranjo Zolotov, M., Oliveira, T., Casteleyn, S.: E-participation adoption models research in the last 17 years: a weight and meta-analytical review (2018). https://doi.org/10.1016/j.chb.2017.12.031
51. Indregard, S.: Det som skiller oss (og ikke) (2019)
52. Mcallister, I.: The personalization of politics. In: The Oxford Handbook of Political Behavior. Oxford University Press (2007)

Ethical Governance of Emerging Digital Technologies in the Public Sector
Insights from Dutch Digital Ethics Commissions

Antonia Sattlegger

Delft University of Technology, 2628 CD Delft, The Netherlands
a.s.sattlegger@tudelft.nl

Abstract. Emerging digital technologies, such as algorithms and machine learning, offer transformative opportunities to public sector organizations but also pose ethical risks and dilemmas. Public sector organizations adopting these technologies need to govern their ethical implications. This research provides insights into the emerging phenomena of digital ethics commissions within Dutch public sector organizations. Composed of external experts on ethics, technology, and governance, these commissions are meant to reflect, advise, and, in some cases, assess the ethical design and use of emerging digital technology and the governance thereof. Through interviews and document analysis, this research explores the motivations, intentions, and perceptions guiding these commissions. Our preliminary findings suggest that, while digital ethics commissions are meant to convey legitimacy, they can also provide external knowledge, open the organizations for reflection from and with society, and contribute to different types of control. In establishing these commissions, government organizations need to balance the formalization of digital ethics governance with the need for a collaborative and reflective ethical practice conducive to (organizational) learning and ethics as contextual practice. This research contributes empirical insights into how public sector organizations address ethical challenges from emerging digital technologies, offering valuable implications for both practitioners and scholars in the field of public administration.

Keywords: Emerging Digital Technologies · Digital Ethics Commissions · AI Ethics · Digital Governance

1 Introduction

Public sector organizations increasingly experiment with and employ emerging digital technologies, particularly those that are data-intensive, such as algorithms and artificial intelligence, to enhance service delivery, improve efficiency, and foster government responsiveness. However, alongside these innovative and transformative opportunities come significant ethical dilemmas and risks. The unethical design and use of emerging digital technologies by public sector organizations not only risks harming individuals and marginalized groups but potentially undermines trust in government within society at large. The deployment of opaque predictive analytics in welfare service distribution may lead to discriminatory decision-making, limiting human discretion and avenues for redress [1–3]. In particular, core public values, public accountability, and transparency are found to be under threat by the use of algorithms [4].

The ethical design and use of emerging technologies have received more attention, both in practice and research [5–7]. This has resulted in a wealth of design and principle-based ethics codes, guidelines, frameworks, and assessments for emerging technologies, particularly when framed as AI. However, there is a discrepancy between principles and practice. Their impact on the actual design and deployment of AI technologies remains limited [8, 9]. Guidelines primarily focus on theoretical underpinnings rather than offering practical guidance on implementation [10, 11], but remaining "deontological tick-box exercises" [12]. As a result, there's a growing acknowledgment of the imperative for concrete ethical and socio-legal governance mechanisms to ensure the ethical design and use of emerging technologies by public sector organizations [13]. However, there remains a scarcity of empirical research examining how public sector organizations translate these abstract ethical principles into actionable practices and effectively address ethics in governance practices [14].

We find that public sector organizations in the Netherlands are increasingly establishing digital ethics commissions to facilitate the ethical design and use of emerging digital technologies. These commissions serve as forums composed of external experts in ethics, technology, and governance, tasked with advising on specific cases, themes, and governance issues related to the ethical design, use, and governance of emerging digital technologies. By providing external advice and guidance, these commissions are meant to support public sector organizations in navigating the ethical complexities of emerging digital technologies.

In other domains, commissions dedicated to ethics have been around for longer and, seemingly, stood the test of time. In the healthcare domain, healthcare ethics committees and research ethics committees have been established since 1970. Ethics committees in healthcare first saw a surge in the United States. Highly publicized cases, such as ending life-saving medical treatment of infants born with Down syndrome or clinically dead patients, sparked societal debates around the ethical implications of modern medicine. Subsequent legal cases highlighted the ethical, legal, and professional limitations of courts to address these ethical dilemmas in practice [15]. Emerging technologies have accelerated the possibilities of modern medicine, and these innovative opportunities have given rise to ethical dilemmas. The ambiguous and uncertain nature of decision-making in patient care was exacerbated by (technological) advancements in modern medicine, coupled with societal changes regarding the increasing recognition of individual patient rights and autonomy and the pluralistic clinical and societal context of healthcare institutions [15]. More recent techno-medical innovations, such as genome sequencing and editing, have further exacerbated medical possibilities and respective ethical dilemmas. Medical ethics commissions are to address these uncertainties in research and practice by providing ethical advice inpatient cases and research projects, developing and revising ethical guidelines in applying new technologies, tools, and processes, and providing ethical training to staff. As such, in the medical domain, ethics commissions are part of broader governance to reduce uncertainties and address ambiguities in the design and delivery of responsible health care [16].

Public sector organizations face similar uncertainties and ambiguities. Emerging digital technologies, such as algorithms, predictive analytics, or large language models, may have a transformative impact on public service delivery by public sector organizations. In this transition period, public sector organizations lack the laws and norms as well as experience in governing emerging technologies and their ethical implications. One may argue that digital ethics commissions could be an instrument to govern the design and use of emerging technologies. Grimmelikhuijsen & Meijer [17] emphasize that safeguarding the legitimacy of algorithmic government requires a multiplicity of different institutional arrangements. Such strategies include the strengthening of civic participation and critical thinkers in the design, development, and monitoring of algorithmic systems, which allows "the voices and values of non-technical experts to be given a place in the development of algorithmic systems." (p. 240). Krijger et al. [18] attribute "the introduction of ethics review boards, codes of ethics, and the engagement of stakeholders in data science processes" as a means to integrate and "promote a culture of ethics and creating awareness for the ethical aspect" throughout the organization, and therefore advanced AI ethics maturity.

Yet, given the heightened public scrutiny of the public sector deployment of emerging digital technologies, PSOs may be tempted to legitimize their actions by a symbolic digital ethics commission. "The setting up of advisory groups that may be powerless or insufficiently critical," Floridi [19] argues, is indicative of ethics bluewashing. An unethical practice which Floridi [19] defines as "the malpractice of making unsubstantiated or misleading claims about, or implementing superficial measures in favor of, the ethical values and benefits of digital processes, products, services, or other solutions in order to appear more digitally ethical than one is" (p. 187). These attempts, whether intentionally or not, "mask and leave unchanged any behavior that ought to be improved" (p. 187). Ethical principles are decoupled from daily practices and respective outcomes, namely the responsible design of emerging technologies [20]. Therefore, in this research, we address the question of how digital ethics commissions are implemented to contribute to the responsible design and use of emerging digital technologies at public sector organizations.

In this explorative research design, we first conduct a literature review on (digital) ethics commissions, examining their conceptualization and role as translational tools in other domains of applied ethics, particularly corporate AI ethics and biomedical ethics. Second, we detail our methodology, which involves semi-structured interviews, document analysis, and the author's reflections, to investigate digital ethics commissions in the Netherlands. Fourthly, we present our study's empirical findings on the perceived needs and motivations for ethics commissions. We provide a definition of digital ethics commissions and distinguish different types of digital ethics commissions according to their impact on the responsible design of emerging digital technologies. Lastly, we discuss our findings, analyzing the contributions and challenges of digital ethics commissions and offering insights for future research and limitations of our study. Our paper aims to contribute to the scholarly discourse on the design, functioning, and implications of digital ethics commissions in public sector organizations, offering empirical insights for addressing the governance of ethical implications of emerging digital technologies to both academia and practice.

2 Literature Review on Digital Ethics Commissions

Digital ethics commissions could serve as a translational tool between ethical principles and practice and facilitate the ethical design of emerging digital technologies at public sector organizations. In the biomedical domain, similar ethics boards are established in practice. In the (corporate) AI ethics domain, such commissions have only recently appeared. Although the latter has built on the tradition of the former, both domains ascribe different functions and roles to the commissions and, thus, build on different responsibility attributions in applied ethics.

In the biomedical domain, these commissions have been around for a long time. Although they do not address digital emerging technologies specifically, technological innovations, along with societal shifts towards recognizing individual patient rights and autonomy within a pluralistic clinical and societal context of healthcare institutions [15], have given rise to ethical dilemmas, making these commissions seemingly more relevant. Recent techno-medical innovations, such as genome sequencing and editing, have further expanded medical possibilities and heightened associated ethical dilemmas. Two types of ethics commissions can be distinguished [21]: research ethics committees (RECs) and healthcare ethics committees (HECs). RECs have a specified (legal) mandate to review and oversee (medical) research involving human subjects. These committees evaluate the ethical aspects of research protocols, ensuring that the rights, safety, and well-being of the research participants are protected. They assess factors such as informed consent procedures, assessment of potential risks and benefits, data privacy, and compliance with ethical guidelines. These types of research committees have been established at universities and other research organizations to oversee research involving human subjects beyond the medical domain. Healthcare ethics committees (HECs), also known as clinical ethics committees and bioethics committees, are typically established at healthcare institutions involved with patient care. They provide guidance on moral issues relating to patient care, such as moral dilemmas in palliative care. Generally, three functions can be distinguished [15, 16, 22]: (1) ethical case analysis, consultation, and conflict resolution; (2) the development and revision of ethics policies and guidelines, such as patient care protocols; and (3) the education of members, hospital staff, and patients about ethical issues. Despite their maturity, the empirical evidence of their effectiveness is limited and inconclusive [16]. The evaluation of such committees largely relies on subjective measures, such as perceptions of goals attainment, satisfaction and perceived helpfulness of their contributions, particularly, the feeling of relief in moral distress through enhanced understanding and responsibility alleviation. Preliminary success factors are a balance between being embedded in the organization and maintaining a critical independence. Such balance is helped by the multidisciplinary composition of the committees, particularly external expertise in bioethics. Furter, these commissions must stimulate collaboration between staff, patients and the committee to stimulate ethical decision-making [16].

In the (corporate) AI ethics literature, digital ethics commissions, also termed AI ethics boards [23], ethics councils [24], ethics advisory board [11] or ethics review boards [25] haven made a recent appearance. In reviewing academic and practitioner governance approaches to foster ethical AI, Prem [24] names ethics councils and boards as good practice to provide an infrastructure and communities supporting ethical systems design. Yet, there is little empirical research conducted on these entities [24, 25]. Stahl et al. [25] find that public organizations and private companies apply but a fraction of mitigation strategies discussed in the literature (p. 33), including ethics review boards. Interestingly, they find that ethics review boards feature strongly as theoretical means to increase organizational awareness and reflection. They may enable the engagement with external stakeholder and the reflection on internal processes (p. 34). Similarly, Tiell [26] argues that "committee-based governance" at AI companies can provide valuable insights and feedback to designers, engineers, and executive teams on how the possible impact of AI applications in practice. Schuett et al. [23, 27] propose AI ethics boards as improvements to the corporate risk governance of AI companies, including medium-sized research labs and big tech companies. The authors [25] define an ethics board as "collective body intended to promote an organization's ethical behavior" (p. 1), particularly the reduction of societal risks. Following these authors, the AI ethics board could be attributed several responsibilities to control societal risks: advising the board of directors, overseeing model releases and publications, supporting risk assessments, reviewing the company's risk management practices, interpreting AI ethics principles or serving as contact points for whistleblowers (p. 2). The authors discuss additional design choices, such as the structure of the commission as external or internal body, it's composition, decision making process and mandate (p. 14) Morley et al. [11] argue for an "independent multi-disciplinary ethics advisory board" (p. 251) to be a crucial in providing the core "infraethics" (p. 251). In serving as a translational tool, it addresses above mentioned disconnect between ethical AI principles and the practical design of AI systems. Three tasks can be attributed to the independent advisory board [11]: First, the ethics board is to be instrumental in developing a principle-based ethics code in negotiation with those impacted by the AI system. Second, the board develops a process for the validation, verification, and evaluation of algorithmic designs. Rather than operationalizing the ethics code as "an end-goal that can be objectively achieved, observed, quantified, or compared" (p. 246), the ethics board supports the contextual interpretation of principles and provides further translational tools. Third, the ethics board conducts regular audits of the developed AI systems, the design processes, and the companies ethical conduct at large. Particularly the function of independent oversight requires, according to the authors (2021) a balance between devolved and centralized responsibility attribution in the ethical AI governance through a multi-agent system. The need for algorithmic audits has been emphasized widely emphasized. Raji et al. [28] suggest "an internal ethics review board that includes a diversity of voices should review proposed projects and document its view. (...) The purpose of an ethics review board for AI systems includes safeguarding human rights, safety, and well-being of those potentially impacted." (p. 39). Yet, given the lack of accountability and transparency, internal algorithmic audits may not provide the necessary independent assurance and external expertise [29]. Therefore, Morley et al. [11] emphasize the independence of the advisory board providing external interdisciplinary expertise and audit. Moreover, to prevent the externalization of responsibility for ethical AI, responsibility for the ethical design must be clearly distribute among the independent board and the internal company employees.

Generally, in the (corporate) AI literature digital ethics commissions are implemented as part of the broader formal risk management framework. They are attributed oversight and advisory responsibilities and report to the governing body. This role requires the independence, and thus, distance, from the primary operational processes in the organization. A commission that serves as and contributes to the development of translational tools, such as the development of a principle-based ethics code, remains a top-down, principled approach. This conceptualization differs from the understanding of healthcare ethics commissions in the biomedical domain. As discussed above, the primary responsibility of ethics commissions is to review and advise on specific cases, either patients or research projects. Ethics commissions, often composed by operational staff, such as doctors or researchers, interact with and advise those responsible for daily operations.

3 Methodology

For this research, digital ethics commissions are defined as structured forums advising on the design and use of emerging digital technologies. We included digital ethics commissions if they fulfilled the following criteria: (1) they were an autonomous body with a structured way of working, indicated by, for example, a memorandum or charter, (2) they advised, reviewed or provided oversight on the design and use of emerging digital technologies, (3) by public sector organizations, such as local, regional, and national government and executive agencies. A total of 21 commissions were identified across executive agencies, regional, and local government organizations through a snowball search. Subsequently, we were able to include 15 of these commissions in our data collection. Table 1 provides an overview of the digital ethics commissions included in this research. Multiple commissions were in the process of formation or early implementation stages. The data collection included semi-structured interviews with facilitators of digital ethics commissions, ethics experts and auditors from within the organization, as well as external commission members. The interviews focused on three areas: First, the process and motivation for establishing a commission, its drivers and the needs and objectives the commission was intended to address. Second, the design choices that were made, such as its the composition of the commission, the decision-making process, the mandate and responsibilities. Third, the interview covered questions regarding the embedding of the commission in the exiting digital ethics governance and the maturity of said governance. Finally, the interview covered questions on the perceived impact of the commissions, as well as risks and weaknesses. The interviews lasted from 40 to 80 min. The interviews were recorded, transcribed and coded in an open-coding approach. Additional data collection included the gathering of internal and publicly available documents such as official reports, evaluations, and the commission reports on specific cases. These documents provided valuable insights into the activities and decision-making processes of the commissions.

A limitation of our data collection is that we interviewed individuals directly involved with the commissions, rather than those within the organizations responsible for designing the emerging technologies. As a result, our assessment of the impact of these commissions relies on the intentions and perceptions of the interviewees. Furthermore, given

Table 1. Overview Digital Ethics Commissions.

Code	Entity	Type	Established	Objective	Focus/ scope of technology	Mandate	Composition	Example cases
C1	Municipality	Sounding board	2022	Reflection on questions posed by the council, the board, the administrative apparatus, explicitly not an advisory body.	Innovative use of data and new technologies.	solicited and unsolicited reflection, non-binding	5 external members	Design and use of dashboards for decision-making, use of cameras and image recognition
C2	Municipality	Sounding board	2022	Case-specific reflection on ethical dimensions and dilemmas	Digital ethics	solicited and unsolicited reflection, non-binding	5 external members	Predictive analytics application
C3	Municipality	Sounding-board	2022	Case-specific reflection on ethical dimensions and dilemmas	Innovative technologies	solicited and unsolicited reflection, non-binding	5 external members	Data sharing for the proactive identification of youth with multiple issues. Data aggregation application on neighborhoods to support public investment decisions, Sensors in public spaces, city surveillance cameras, corona tracing app.
C4	Municipality	Sounding-board	2021	Case-specific reflection on ethical dimensions and dilemmas	Digital and technological applications involving the collection and processing of data	solicited and unsolicited reflection, non-binding	7 citizen selected on diversity	Automated scan of licenses and parking tickets. Data aggregation in digital twin. Ethical framework for data applications and algorithms.
C5	Municipality	Sounding-board	2019	Citizen-participation and reflection on cases and specific questions	Privacy, digital security, and inclusivity	solicited and unsolicited reflection, non-binding	Self-selected group of citizen	3D mapping via satellite images, COVID-19 Crowd Density Monitoring App

(continued)

Table 1. (*continued*)

C6	Regional Sounding-board 2023	Theme-based reflection on ethical dimensions and dilemmas	Digital technology, algorithms, data applications	solicited and unsolicited reflection, non-binding	7 external members	Advice on the responsible use of ChatGPT
C7	Executive agency Sounding-board 2022	Theme-based reflection on ethical dimensions and dilemmas	New technologies	solicited reflection, non-binding	10 external members	Reflection on topics, such as ethical framework for facial recognition technology, and ethical framework for internet surveillance
C8	Municipality Advisory Committee 2023	Case-specific advice on ethical dimensions and dilemmas, and advice on the development of a value framework	Data and technology	solicited and unsolicited reflection, non-binding	5 external members	Just started
C9	Municipality Advisory Committee 2022	Case-specific advice on ethical dimensions and dilemmas, with particular focus on the governance questions	The use of algorithms, particularly algorithmic governance	solicited and unsolicited reflection, non-binding	9 external members	Facial recognition upon registration (procured software), Register for sensors in public spaces
C10	Municipality Advisory Committee 2018	Case-specific advice on ethical dimensions and dilemmas	Technology and innovation	solicited and unsolicited reflection, non-binding	7 external members	surveillance in high-risk areas within the municipality, app to gain better insight into night-time street harassment, advice on Principles of the Digital Society
C11	National Advisory Committee 2023	Case-specific advice on ethical dimensions and dilemmas	Data analysis, algorithms, risk models, artificial intelligence	solicited and unsolicited reflection, non-binding	11 external members	Just started
C12	Municipality Supervision Committee 2020	Case-specific advice on ethical desirability	Complex and/or politically sensitive personal data processing technology	solicited and unsolicited reflection, formal comply-or-explain, official responsible reports deviation to council and	7 external members	Risk model for welafre support applications, algorithmic bias analysis

(*continued*)

Table 1. (*continued*)

C13	Executive agency	Supervision	2022	Case-specific ethical judgement	Data applications	solicited, comply-or-explain	Multiple internal members	Provision of data and data protection
C14	Executive agency	Supervision	2021	Case-specific ethical judgement	Data research projects using personal data	solicited, binding	Multiple internal members	Biases in data research projects
C15	Executive agency	Advisory/Supervision	2022	Case-specific advice on ethical dimensions and dilemmas	Personal data analysis, algorithms, predictive analytics, artificial intelligence with impact on citizens.	solicited and unsolicited reflection, non-binding	7 internal and 2 external members	Fraud prediction algorithm, proactive service based on vulnerability prediction algorithm

that digital ethics commissions are a relatively recent phenomenon, with most established within the last two years, their mid- and long-term impact remains to be assessed. A potential limitation arises from the involvement of two of the authors with different commissions at an executive agency. While these commissions were not formally included in the research, the authors' experiences may have influenced their reflections on ethics commissions. To mitigate this risk, shared reflection and critical analysis were employed throughout the research process. This research was motivated and informed by the authors practical experience with digital ethics commissions in the Netherlands. The first author is involved as member of a digital ethics work group facilitating an ethics commission at an executive agency (C15). The second author is active member of two ethics commissions (C6 & C9). This involvement and the experience in working with ethics commissions from different perspectives is a valuable insight. This personal connection also introduces a personal bias. We seek to mitigate this risk by shared reflection and critical analysis through comparing our experiences to other commissions.

4 Results

In the following section, we present our empirical results by addressing three key questions: First, we identify the perceived challenges and needs in the adoption of emerging digital technologies which motivated the establishment of these commissions. Second, we analyze how these commissions are meant to address these needs. By deducing shared characteristics of these commissions, we arrive at a definition of digital ethics commissions as employed by Dutch public sector organizations. Third, we explore the different ways digital ethics commissions are intended to contribute to the responsible design and use of emerging digital technologies, culminating in a typology of their various designs.

4.1 What are Digital Ethics Commissions?

To address these needs, public sector organizations have, or are in the process of, establishing digital ethics commissions. We find that digital ethics commissions can be defined as structured forums advising on the design and use of emerging digital technologies. Table 1 provides an overview of the commissions we identified in the Netherlands. While the design of commissions varies, we find that they generally share three characteristics.

Scope. Digital ethics commissions address ethical challenges and dilemmas related to the adoption and governance of emerging digital technologies. Particularly sounding boards employ a broader scope, including technologies such as open data platforms, digital twins, geospatial technologies and smart city technology, such as drones, surveillance cameras, or crowd control. Supervision commissions adopt a narrower scope, typically addressing algorithms and artificial intelligence. The working definition of these technologies is subject to discussion and tends to become more encompassing with the increased maturity of the commission.

Structured Forums. These commissions are composed of a purposefully selected group of members, have a structured way of working and a formal mandate. This organization gives the commissions a certain degree of agency and allows them to engage in a shared learning and development process with the organization, differing from the ad-hoc participation of citizen or consultation of experts. Commissions generally have a specified formal mandate to provide advice on ethical issues, though the strength of this mandate varies, as we will discuss below. This advice is usually provided in a written report which reflects the reflection and assessment of the commission members. Generally, commissions take a concrete case as their starting point, although some are also reflecting on broader questions. In those commissions who work with a case-based approach and prepare specific questions to deliberate with the commission, both the facilitators and the members seem to be more positive about the useability and impact of the advice. As one member argues: "it is often the application of the technology in the context that really provides depth. So what values are really at stake here? So that was really looking at it from a case-by-case perspective, which also makes the conversation more valuable and ultimately the yield as well." (C3F1). Another commission member agrees "We also looked at how they deal with an algorithm register? How do they ensure they have enough overview and which algorithms are all within the Municipality? The cases are still a kind of entry point to discuss broader policy. So ultimately, the cases are not the core for us. But my experience is that if you only talk about policy in abstract terms, then you actually have no idea how that works out in concrete practices. Practice is always also an entry point for a broader kind of consideration." (C9M1).

Degree of Independence. The independence of the commission is repeatably emphasized by the interviewees and the documentation of the commissions. Though degrees may vary, this independence lies in the external composition of the commission, its positioning as autonomous body and its mandate to give solicited as well as unsolicited advice. Although the ability to provide unsolicited advice seem to play a rather ideational role, as we could find only two instances of unsolicited advice. The commissions are external by selecting members from outside the organization. Commissions are composed of external experts with a background in (applied) ethics, computer science, public administration, business analysts, and legal experts. Particularly university professors are perceived as conveying a high degree of independence. Other commissions are composed of citizen with or without specific knowledge in the domain. Both compositions are meant to provide an external and independent view on the organization's inner workings, as "view from the outside" (C4F1) by "people from other worlds. With a different perspective on things." (C9F1). Other commissions consist of internal members or a mix of internal and external members. In these cases, interviewees emphasize the independence of the commission's advice by highlighting the autonomy granted by its mandate and self-governing capabilities.

4.2 Which Needs are Digital Ethics Commissions Meant to Address?

Digital ethics commissions are generally meant to address four interrelated challenges which arise with emerging digital technologies: the perceived complexity, uncertainty, and ambiguity related to the adoption of emerging digital technologies, and socio-political pressures, such as the lack of societal trust in the use of emerging digital technologies, particularly algorithms and artificial intelligence.

Complexity. Interviewees experience various types of complexity, such as technological, epistemic, and organizational complexity. For one, complexities relating to the digital technology and its use. This is exemplified by the discussions and the perceived need for a shared understanding of (technical) concepts, such as algorithms or artificial intelligence. This complexity is related to the "interweaving of technology and ethics and other governance issues. These questions involve intertwining backgrounds, knowledge of technology, but also knowledge of law or governance. Or ethics." (C9M1). In their perception, public organizations are particularly unequipped to deal with such interconnected complexities, which requires cross-silo thinking. This "interweaving" requires the involvement of multiple organizational entities such as the FG, the CIO, the CDO, in some cases an ethics officer. This creates a "complex field of forces surrounding it" (C9M1).

Uncertainty. Interviewees also express uncertainty regarding the emerging and transformative character of these technologies. For one, technology is rapidly evolving, as well as the possibilities and use-cases it affords. For another, interviewees express uncertainty about the possible impact on citizens and society. The transformative impact of these technologies is expected to be great, and so are the associated risks – "you still have a kind of yes unknown territory where the risks are estimated to be high." (C9M1). Due to the novelty and emerging character of these technologies, there is a lack of experience, norms and regulation governing them. One interviewee summarizes "The laws and regulations are lagging behind, so we have to do other things to ensure that it is used in a responsible manner." (C11M2). While this delay is perceived in terms of time, stricter regulation is also not always desirable, as one participant argues: "And you always have that consideration that you don't want to be too restrictive because then you might hinder certain innovations. Sometimes it's just very complicated to see when new technology comes into society and what the negative effects of that actually are." (C10M1).

Ambiguity. Interviewees also express ambiguities in the application of these technologies and the opportunities they afford. This is exemplified by an often-raised dilemma between what is perceived as desirable and what is legally possible in the context of using personal data for bias testing. Interviewees perceive these value conflicts to be particularly pressing in the context of public organizations, as they not only have to realize moral values through their design artifacts, such as fairness, but also public values through trough the artifact, the process and organization. This ambiguity is often not reflected in ethical design principles, C7M1 argues.

Transition Period. Facing these challenges, many interviewees recognize that public sector organizations find themselves in a transition period. This transition period is characterized by the above-mentioned emerging challenges. Together with a developing governance structure for emerging technologies, and particularly the governance of ethical aspects. When asked to assess their organizations maturity according to the AI ethics maturity framework by (Krijger et al., 2023), interviewees assess the maturity as generally low, particularly on the level of "orientation on frameworks, guidelines/principles, trainings on data science ethics takes place in teams". Overall, the organizations find themselves in a "search" (C2F1) or "learning process" (C8F1), and they "are just not there, yet" (C3F1) to effectively address the ethical implications of emerging digital technologies.

Lack of Legitimacy and Trust. The use of predictive algorithms, as one particular type of emerging digital technologies, by public sector organizations in the Netherlands is under increased public scrutiny, especially in the social welfare domain. A prominent case involved the Dutch tax authorities using a predictive algorithm to wrongfully accuse childcare benefits recipients, primarily from minority backgrounds, of fraud. This has heightened public sensitivity and generated distrust towards the government's use of algorithms. All but two interviewees mentioned this case to illustrate the importance of digital ethics (commissions) at one point of the interview. Public sector organizations are perceived to be standing "under a magnifying glass" (C3F1) – doing "anything with algorithms, you are already on the backfoot" (C15M1). The increased political sensitivity and distrust contributes to a need of legitimization of the design of algorithms. Digital ethics commissions have been established largely at the initiative of political-administrative leadership. Respectively, the need for digital ethics commissions is not always shared. As one interviewee remarks, "I wasn't waiting for it (the digital ethics commission)," states an interviewee, "we were already doing well with an internal working group." (C8F1). While there may be various means to legitimize the design of algorithms, digital ethics commissions are particularly widely applied and discussed in the landscape of government organizations. Multiple interviewees raise mimetic and normative isomorphism [30] as drivers for the establishment of digital ethics commissions across government organizations. "Every municipality that wants to be taken seriously now needs an ethics commission. Municipality [X] has one, so does municipality [Y], then we need one too. That is just how it goes. It's has become a trend." (Interviewee).

4.3 What are Different Design Choices of Digital Ethics Commissions that are Made to Address These Challenges?

Government organizations design digital ethics commissions differently to address these needs. The need to address the internal challenges related to the complexity, uncertainty and ambiguity and the external socio-political pressures arising with the adoption of emerging digital can be either addressed along an important continuum. On the one end, the formalization and assurance of digital ethics governance through the digital ethics commissions. On the other end, the facilitation of external ethical reflection and collective engagement in promoting and developing ethical development. We find that commissions develop from their initial intention and throughout their implementation and may be understood as different types. Generally, we find that three types of ethical commissions can be defined along this line:

Sounding board – A sounding board provides ethical reflections on topics and governance put forward by the facilitator or working group.
Advisory committee – An advisory committee provides actionable recommendations on dealing with ethical dilemmas in specific projects and governance aspects. An advice can have a formal or informal comply-or-explain mandate.
Supervision commission – A supervision commission provides a judgement as to whether it is perceived as ethical to proceed with a project.

While there is variance both across the commissions as well as in their development, we find the following characteristics to be indicative for these different types of commissions – and the respective understanding of governing and implementing digital ethics.

Mandate – Three types of mandates can be differentiated: First, a reflection by sounding boards, provides identifications and reflections on ethical dilemmas, or "the uncovering of certain tensions" (C11M2). Reflections are formulated to avoid recommendations in terms of actionable guidance or course for action. As one commission member elaborates, "(…) Our reports are a contribution to the political debate and are not final judgment. It is rather a starting point for a kind of joint reflection. Second, an advice by advisory boards, which provides recommendations on whether or not a technology or certain course of action is ethically desirable, as well as assessments on moral values and dilemmas. To arrive at an advice, commissions often, but not always, use an assessment framework, such as a catalogue of organizational core values. An advice is non-binding and rarely supported by a formal comply-or-explain procedure. Third, a judgement by a supervision commission, which entails a clear directive as to whether to proceed or not with a certain project. There is only one organization whose commissions judgements has a strong comply-or-explain mandate.

Addressee – Closely related to the mandate, we find that digital ethics commissions vary as to the actors they aim to address, engage with and impact. Supervision commissions address case owners, such as the designers and developers with their ethical judgments. While cumulative reports are extended to administrative leadership, case owners are responsible for approaching the commission and implementing the ethical judgment. Sounding boards seek to address a broader audience, particularly political-administrative leadership. One commission member understands the commission as "an attempt at democratizing ethics. (...) On the one hand, introducing ethical arguments in a flexible way into the democratic debate in the City Council. So, enabling councilors to have as broad an ethical discourse as possible about new technology. (...) And on the other hand, empowering citizens to ask those ethical questions in a much broader way themselves. And I think that's really the strength of such an approach." (C3M2). Advisory committees generally address both case owners and administrative leadership. Interestingly, despite their varying mandates and addressees, digital ethics commissions generally extend their reflections as written reports. Few commissions engage or experiment with different forms which could be more inducive to shared reflection and learning. Two advisory boards emphasize the importance of shared sessions between the commission and internal working groups. Another, supervision commission organizes annual panels as means of interacting with political-administrative leadership.

(Dis-)integration in the organization – Lastly, the integration and disintegration of the ethics commission form the organization is indicative for the different types of commissions. Supervision commissions are most closely integrated and formalized in the organizational processes and structures. Often, they are understood to be part of the organizational risk management or are connected to existing privacy governance. Sounding boards are the most removed from the organization. This is exemplified by how cases are referred to the commissions. Particularly, supervision boards, and, to increasingly advisory boards, have a formal process of referring cases to commissions. In few instances organizations have established a risk scan, which indicates when case owners are to engage with the commission. In other cases, cases are escalated through the internal working group to the commission, such as, if a certain dilemma is perceived to be sufficiently "challenging", "interesting" or "representative" by the facilitator. Again, advisory boards fall in between and, therefore, interviewees at times express a tension between the independent mandate of the commission and the embedding in the organization.

5   Discussion, Conclusion and Limitations

In this research we explore the question as to how digital ethics commissions are implemented to contribute to the responsible design and use of emerging digital technologies at public sector organizations. We find that digital ethics commissions are addressing both design and use challenges related to the degree of complexity, uncertainty and ambiguity which emerging digital technologies introduce, as well as a need for assurance and oversight related to the perceived lack of legitimacy and public trust in the governments use of emerging digital technologies, particularly algorithms and artificial intelligence. Public sector organizations address these needs differently through varying implementations of digital ethics commissions – sounding boards, advisory committees and supervision commission. Digital ethics commissions may address these needs by providing external knowledge and reflection. They may also serve as assurance and oversight body, which relates to the functions that commissions are attributed in biomedical ethics and (corporate) AI ethics. Digital ethics commissions can also be indicative for a balance public sector organizations are seeking between the formalization digital ethics governance instruments, procedures and practices in bureaucratic organizations, on the one hand, and facilitating collaborative ethical reflection and organizational learning on the other hand.

While we have analyzed the motivations, intentions, and perceptions behind the establishment of these commissions, we have not evaluated the actual impact on design and use of emerging digital technologies by public sector organizations. This limitation should be addressed through further research into the actual impact of ethics commissions in on actual organizational practices and their materialization in the design of technology. Such research should include those addressed by the commissions, particularly project owner and managers, and administrative as well as political leadership. This is another limitation of this study. Our interviews were limited to facilitators of the commissions. Though these individuals are most knowledgeable in the workings of such commissions, they personal stake in the commissions could introduce a bias. A bias could also be introduced through the authors involvement in the work of ethics commissions, as we have previously discussed.

This paper offers preliminary empirical insights into how public sector organizations address the ethical challenges posed by emerging digital technologies through digital ethics commissions. By analyzing the motivations, intentions, and perceptions guiding the establishment of digital ethics commissions, it illustrates how government organizations seek to balance the need for organizational digital ethics maturity and formalization, on the one hand, and the need for shared reflection and learning in a transition period.

References

1. Margetts, H.: Rethinking AI for good governance. Daedalus **151**, 360–371 (2022). https://doi.org/10.2307/48662048
2. Peeters, R., Widlak, A.C.: Administrative exclusion in the infrastructure-level bureaucracy: the case of the Dutch daycare benefit scandal. Public Adm. Rev. **83**, 863–877 (2023). https://doi.org/10.1111/PUAR.13615
3. Peeters, R., Widlak, A.: The digital cage: administrative exclusion through information architecture – the case of the Dutch civil registry's master data management system. Gov. Inf. Q. **35**, 175–183 (2018). https://doi.org/10.1016/j.giq.2018.02.003
4. Busuioc, M.: Accountable artificial intelligence: holding algorithms to account. Public Adm. Rev. **81**, 825–836 (2021). https://doi.org/10.1111/puar.13293
5. Tsamados, A., Aggarwal, N., Cowls, J., et al.: The ethics of algorithms: key problems and solutions. AI Soc. **37**, 215–230 (2022). https://doi.org/10.1007/s00146-021-01154-8
6. Buijsman, S., Klenk, M., van den Hoven, J.: Ethics of Artificial Intelligence
7. van de Poel, I.: Embedding values in Artificial Intelligence (AI) systems. Minds Mach. (Dordr) **30**, 385–409 (2020). https://doi.org/10.1007/S11023-020-09537-4
8. Jobin, A., Ienca, M., Vayena, E.: Artificial Intelligence: the global landscape of ethics guidelines. Nat. Mach. Intell. **1**, 389–399 (2019)
9. McNamara, A., Smith, J., Murphy-Hill, E.: Does ACM's code of ethics change ethical decision making in software development. In: Proceedings of the 26th ACM Joint European Software Engineering Conference and Symposium on the Foundations of Software Engineering (ESEC/FSE 2018), Lake Buena Vista, FL, USA, 4–9 November 2018. ACM, New York (2018)
10. Morley, J., Floridi, L., Kinsey, L., Elhalal, A.: From what to how: an initial review of publicly available AI ethics tools, methods and research to translate principles into practices. Sci. Eng. Ethics **26**, 2141–2168 (2020). https://doi.org/10.1007/s11948-019-00165-5

11. Morley, J., Elhalal, A., Garcia, F., et al.: Ethics as a service: a pragmatic operationalisation of AI ethics. Minds Mach. (Dordr) **31**, 239–256 (2021). https://doi.org/10.1007/s11023-021-09563-w
12. Hagendorff, T.: The ethics of AI ethics: an evaluation of guidelines. Minds Mach. (Dordr) **30**, 99–120 (2020). https://doi.org/10.1007/s11023-020-09517-8
13. Theodorou, A., Dignum, V.: Towards ethical and socio-legal governance in AI. Nat. Mach. Intell. **2**, 10–12 (2020)
14. Sigfrids, A., Nieminen, M., Leikas, J., Pikkuaho, P.: How should public administrations foster the ethical development and use of artificial intelligence? A review of proposals for developing governance of AI. Front. Hum. Dyn. **4**, 858108 (2022). https://doi.org/10.3389/FHUMD.2022.858108/FULL
15. Aulisio, M.P., Arnold, R.M.: Role of the ethics committee: helping to address value conflicts or uncertainties. Chest **134**, 417–424 (2008). https://doi.org/10.1378/CHEST.08-0136
16. Crico, C., Sanchini, V., Casali, P.G., Pravettoni, G.: Evaluating the effectiveness of clinical ethics committees: a systematic review. Med. Health Care Philos. **24**, 135–151 (2021). https://doi.org/10.1007/S11019-020-09986-9
17. Grimmelikhuijsen, S., Meijer, A.: Legitimacy of algorithmic decision-making: six threats and the need for a calibrated institutional response. Perspect. Public Manag. Gov. **5**, 232–242 (2022). https://doi.org/10.1093/ppmgov/gvac008
18. Krijger, J., Thuis, T., de Ruiter, M., et al.: The AI ethics maturity model: a holistic approach to advancing ethical data science in organizations. AI Ethics (2022). https://doi.org/10.1007/s43681-022-00228-7
19. Floridi, L.: Translating principles into practices of digital ethics: five risks of being unethical. Philos. Technol. **32**, 185–193 (2019). https://doi.org/10.1007/s13347-019-00354-x
20. Meyer, J.W., Rowan, B.: Institutionalized Organizations: Formal Structure as Myth and Ceremony (1977)
21. Steinkamp, N., Gordijn, B., Borovecki, A., et al.: Regulation of healthcare ethics committees in Europe. Med. Health Care Philos. **10**, 461–475 (2007). https://doi.org/10.1007/s11019-007-9054-6
22. Hajibabaee, F., Joolaee, S., Cheraghi, M.A., et al.: Hospital/clinical ethics committees' notion: an overview (2016)
23. Schuett, J., Reuel, A.-K., Carlier, A.: How to design an AI ethics board. AI Ethics (2024). https://doi.org/10.1007/s43681-023-00409-y
24. Prem, E.: From ethical AI frameworks to tools: a review of approaches. AI Ethics (2023). https://doi.org/10.1007/s43681-023-00258-9
25. Stahl, B.C., Antoniou, J., Ryan, M., et al.: Organisational responses to the ethical issues of artificial intelligence. AI Soc. **37**, 23–37 (2022). https://doi.org/10.1007/s00146-021-01148-6
26. Tiell, S.: Create an Ethics Committee to Keep Your AI Initiative in Check. Business Ethics (2019)
27. Schuett, J.: Three lines of defense against risks from AI. AI Soc. (2023). https://doi.org/10.1007/s00146-023-01811-0
28. Raji, I.D., Smart, A., White, R.N., et al.: Closing the AI accountability gap: defining an end-to-end framework for internal algorithmic auditing. In: FAT* 2020 - Proceedings of the 2020 Conference on Fairness, Accountability, and Transparency, pp. 33–44. Association for Computing Machinery, Inc. (2020)
29. Mökander, J., Floridi, L.: Ethics-based auditing to develop trustworthy AI. Minds Mach. **31**, 323–327 (2021). https://doi.org/10.1007/s11023-021-09557-8
30. Dimaggio, P.J., Powell, W.W.: The iron cage revisited: institutional isomorphism and collective rationality in organizational fields. In: The New Economic Sociology, pp. 111–134 (2022). https://doi.org/10.1515/9780691229270-005/PDF

Data Pollution: Definition and Policy Responses

Leonardo Mori(✉), Alizée Francey, and Tobias Mettler

Swiss Graduate School of Public Administration, University of Lausanne,
Chavannes-près-Renens, Switzerland
`leonardojacopo.mori@unil.ch`

Abstract. Data has become one of the strongest drivers of economic growth and innovation. However, this data-driven transformation brings various challenges and harms affecting our lives and environments across different domains and scales. In this article, we define the set of such harms as *data pollution*. Data pollution is a multifaceted phenomenon, entailing different dimensions and complex mechanisms, which we capture in one conceptual model using network thinking and cybernetics. We further analyse the policy landscape to comprehend the awareness level and responses to this phenomenon.

Keywords: Data pollution · conceptualisation · model · cybernetics

1 Introduction

The surge in data, often termed the *data deluge*, is a direct outcome of substantial advancements in various domains, including the widespread adoption of smartphones, the ubiquity of social media platforms, the pervasive integration of the Internet of Things (IoT), among others [1, 2]. These global trends generate more data than ever, and by 2025, there will be more than 175 zettabytes of data, reflecting a fivefold growth from 2018 to 2025 [3, 4]. This significant growth in data volume and the rates at which data are generated make data the lifeblood of the economy and a driver of innovation and societal progress, notably through the progressive extension of Artificial Intelligence (AI) use, which itself necessitates the analysis of extensive volumes of data [5].

Within this context, the *data economy* emerged as a catch-all term covering all aspects related to the generation, collection, storage, processing, sharing, analysis, and use of data facilitated by digital technologies [4]. Data are thus considered a valuable asset, as most economic activities may depend on data within a few years [5]. While the value of the data economy of EU27 was almost €325 billion in 2019, representing 2.6% of the gross domestic product (GDP), predictions foresee that the European Union (EU) data economy will be worth €550 to €829 billion in 2025, representing 4% to 6% of the overall EU GDP [4, 5]. To achieve this, the European Commission supports data sharing through legislation and practical measures, notably by publishing a sequence of directives, strategies, and regulatory acts to set directions for the EU member states [4].

As illustrative examples, critical documents include the General Data Protection Regulation (GDPR) (2018), which sets standards for data privacy; the Open Data Directive (2019), which provides standard rules for government-held data by addressing barriers to the reuse of publicly funded information; the European Data Strategy (2020), which is oriented toward establishing a unified data market that not only bolsters Europe's competitiveness but also fortifies its control over the data; the Data Governance Act (2021), which promotes the availability of data by allowing reuse of some categories of protected public sector data; and the Data Act (2022), which sets rules for the use of data generated by IoT-enabled devices [4]. Moreover, the creation of nine European data spaces aims to facilitate the secure and cost-effective exchange of data across the EU, encompassing both public sector and business data to stimulate the growth of novel data-driven products and services [5, 6]. In addition to promoting data sharing, government at all levels and from all parts of the World have been dedicating remarkable effort to the digitalization of their own operations for now decades [7, 8], themselves largely contributing to the data deluge [9]. This trend is bound to accelerate in the coming years, as a 2022 survey to senior officials of 200 city governments across the World reports that respectively 73% and 49% of respondents identified making real-time decisions from data and making data accessible to the public as priorities for the 5 years to come [10].

As data fuels the new economy by creating endless opportunities, it becomes to this century what oil was to the last one, and it thus pollutes [2, 11, 12]. As an illustrative example, the Shift project, a think tank promoting the transition to a post-carbon economy, has estimated that the proportion of worldwide greenhouse gas emissions attributable to data has risen from 2.5% in 2013 to 3.7% in 2019 [12, 13]. However, harms generated by data expand well over ecological pollution, and following the pivotal role of data in the new economy, this paper argues for the imperative to acknowledge such concomitant deleterious harm; we refer to these as *data pollution*. Since the 1980s, data pollution has been used in different contexts with different meanings, including insufficient quality data [14, 15] but also the external repercussions that emanate from data use such as social and ecological side effects [11, 12]. We argue that these different meanings all emanate from a common cause, being the massive increase in data availability, and interact with each other through cause to effect relationships and ultimately forming a complex problem. This article thus does not propose a new definition of data pollution per se, but rather to relate the different existing meanings through a conceptual model dissecting the dynamics contributing to data pollution. This article defines data pollution as the set of harms generated by any data activity. The conceptual model's primary objective is to develop a comprehensive understanding of data pollution to render these harms perceptible systematically and acknowledged while concurrently elucidating the existing regulatory mechanisms at the EU level. We believe that such an understanding is particularly relevant for governments, both for evaluating the impacts of their own data related initiatives, and to regulate those from the private sector. Finally, we conclude that addressing data pollution with a comprehensive perspective is necessary for the promotion of sustainability, responsible data management, and the protection of individual and societal well-being.

2 Background

2.1 Data Pollution Definitions in the Extant Literature

Data pollution has been used in various pieces of research since the 1980s. Over the decades, it has been employed by different authors from different research disciplines, such as machine learning [16], computer networking [17], AI research [18], and even neuropsychiatry [19]. These disciplines are highly diverse, and the meaning given to data pollution in extant literature is varied and strongly dependent on the context in which it is used, sometimes referring to pollution *of* the data, and other times to pollution *by* the data. Nonetheless, data pollution is often used to designate recurring phenomena with three frequently associated meanings.

The first mention of data pollution dates from 1986 and refers to it as "*the accumulation of all 'contaminations' or 'distortions' which can result from working with data in the information technology field*" ([15], p. 291). Still today, data pollution is often used to refer to data that is of bad quality, untrustworthy, or not predisposed to be used optimally. This broad category includes a spectrum of aspects associated with data pollution, which, although similar, often differ as they may be specific to particular contexts or problematics. Indeed, while older works would be concerned with a general "*contamination of the information supply with incomplete, inconsistent, or incorrect information*" ([14], p. 24), many authors have used the term data pollution to designate more field-specific problematics in recent years. For instance, researchers in machine learning tend to include sample imbalances within data pollution [18, 20]. Another way the meaning of data pollution has been restricted is by referring to it as introducing inaccurate or otherwise unhelpful data into datasets rather than to the exitance of data. Along these lines, publications about network coding systems almost exclusively mention data pollution in the context of attacks, in which "*attackers inject corrupted [data] packets into the network*" ([21], p. 741), while other scholars go as far as defining data pollution based on the unintentionality to introduce errors into the data, in contrast with data poisoning, which refers to voluntary data degradation [19]. In all these cases, polluted data are considered a nuisance because they potentially have adverse effects on the performance of their intended use.

A second meaning of data pollution relates to excessive data production, storage, and publication. Here, data are considered to pollute by their mere existence, as they do not generate benefits and only take up storage space [22]. This happens when data are duplicated or disseminated without there being an interest in it. Not only are these datasets useless, but they can contribute to decreasing the findability of data we would like to use and to a series of unwelcomed consequences, such as privacy violation [22] or even negatively impact our capacity to recognise information from fake news, our concentration and overall well-being [23].

Lastly, recent discussions have emerged over the unwanted effects of data on social environments. According to Ben-Shahar [11], data pollution refers to the harmful effects of the "*exchange of data between giver and taker*" ([11], p. 148). The author argues that while private data leaks are often considered detrimental because of their harm to privacy, their potential damages go well beyond that, as he believes that data production "*creates public harms and destroys public goods*" ([11], p. 106).

2.2 Toward an Overarching Definition of Data Pollution

Although extant literature has identified several meanings and related aspects that can be attributed to data pollution, each was conceptualised and is usually considered in isolation from the others. However, we argue that the various meanings and related aspects of data pollution have a shared origin arising from the increased ability to produce, store and use vast amounts of data and the adjustments of businesses and public organisations' practices under the big data era and its associated data economy. There is thus a gap in the existing literature in understanding data pollution, given that there is currently no overarching view that would effectively contextualise its diverse meanings and related aspects, delineate their origins, and elucidate their complex interactions. Thus, we posit that the interactions between various notions existing in the scientific literature, or even the public debate, and data pollution should be investigated. In pursuit of this, we contend that in addition to those mentioned above, other issues resulting from data activities should be incorporated into the definition of data pollution.

The first of these aspects is the notion of data overload, which in the context of lexicography has been defined as a situation where "the dictionary user gets more data than he or she needs or can deal with during the present consultation and becomes confused and fails to retrieve the necessary information" ([24], p. 397). This idea relates to the over-publication of data mentioned above and the concept of information overload, defined as a situation where "information received becomes a hindrance rather than a help when the information is potentially useful" ([25], p. 249), in some extreme cases even leading to health issues. Thus, we believe that data overload should be included in an overarching definition of data pollution, as we consider it a set of harm resulting from the data. By including data overload, we acknowledge the social adverse effects of excessive exposure and the fact that maintaining and processing excessive volumes of data requires substantial infrastructures and resources, contributing to energy consumption.

The growing mindfulness of environmental challenges is spurring a debate over the ecological pollution related to data. It was estimated that in 2019, digital technology produced 4% of the overall greenhouse gas emissions [26], while data centres and transmission networks alone are the cause of 1% of energy-related greenhouse gas emissions, as a 2023 study by the International Energy Agency (IEA) revealed [27]. In 2022, global data centre electricity consumption accounted for 1 to 1.3% of global final electricity demand, with an annual growth of 20 to 40% in the latest years ([27], p. 2). Given the trend of continuing digitalisation of products and services, carbon emissions, natural resource extraction, production of waste, and other harmful environmental impacts, directly or indirectly, will raise with data-driven infrastructures gaining in economic importance [28]. Such waste is fuelled by the proliferation of electronic devices and their rapid obsolescence. Given that a proportion of the current ecological pollution is due to the set of harm from the data or its use, we believe that it should be integrated into the overarching definition of data pollution to highlight the environmental consequences of our digital lifestyle.

3 Methodology

Given that data pollution exhibits a multifaceted nature, involving diverse dimensions and mechanisms, our decision to utilize network thinking and cybernetics as an approach aims to enhance our understanding of this phenomenon. The central focus of cybernetics is not so much on the structural elements within the system but on their operational dynamics [29]. Cybernetics acknowledges that our understanding of systems relies on our simplified representations or models of those systems and also recognises that simplified representations or models ignore aspects of the system irrelevant to the purpose for which the model is constructed [30]. While data pollution itself is not a complex system, the context from which it stems can be characterised as complex due to the interactive components involved in the emergence of data pollution. We thus understand data pollution within a more extensive system that can be seen as complex. As their name suggests, complex systems are typically hard to understand, but network thinking may ease their comprehension [31]. We thus used network thinking to build our conceptual model on data pollution, which is a collection of nodes and links between these nodes. While the interactions between its different components make the definition and management of data pollution difficult, the feedback loops make detecting and remedying data pollution complex, as actions taken to prevent data pollution may have subsequent effects. This is especially true given that an effect can feed back into its cause in cybernetics. For example, many algorithms use data to propose products and services. If the data are somehow dirty, the algorithms may create bias and inequalities, which may generate dirty data feeding back to the data used by algorithms, leading to increased bias and inequalities, thus creating a vicious circle. Moreover, feedback loops can be positive or negative [30, 32, 33]. The feedback loop is negative if a positive deviation leads to a negative deviation at the following node. For instance, when there is a rise in the volume of data, it reduces search efficiency. Subsequently, this reduction in search efficiency leads to a decline in the retrievable data volume, as only a fraction of the data can be accurately found. The opposite situation, where an increase in the deviation produces further increases, is called a positive feedback loop. For example, a rise in the volume of data will allow for more innovative products and services that will themselves generate data, thus leading to higher data volume. This straightforward method allows us to ascertain whether a given loop will result in stabilisation (indicative of a negative feedback loop) or an unrestrained and escalating process (indicative of a positive feedback loop) [30].

Our research started with the identification of key nodes involved in the genesis and propagation of data pollution. This analytical foundation enabled us to construct a nuanced and systematic representation of the complex interactions shaping the landscape of data pollution. Altogether, the interactions between the nodes facilitated the creation of a visual representation of the system, which we refer to as the conceptual model. We then reviewed and refined our conceptual model by stimulating the analysis of different instances. This process significantly enhanced our conceptual model's precision and depth. Finally, we ensured that the conceptual model adeptly captured the nuances and dynamics of data pollution, which enabled the additional or removal of nodes, the recalibration of the interactions, and the incorporation of feedback loops. By incorporating the role of feedback loops from cybernetics, the model considers the interactions and interdependencies of the nodes. Moreover, in both network thinking and cybernetics, the flow of information is crucial and combining both shows how data pollution flows within and between the nodes. Combining network thinking and cybernetics can lead to a more comprehensive understanding of data pollution and provide valuable tools for managing and mitigating it. The objective of this model is to propose a new conceptualisation of the most discussed forms of data pollution, showing how they stem from a common phenomenon and interrelate. We hope that our model can stimulate discussion and serve as a basis for further work expanding it with other related data harms.

4 Conceptualising Data Pollution

Figure 1 shows our conceptual model, aiming to provide an overview of the complex mechanisms generating data pollution, which we describe in more detail next.

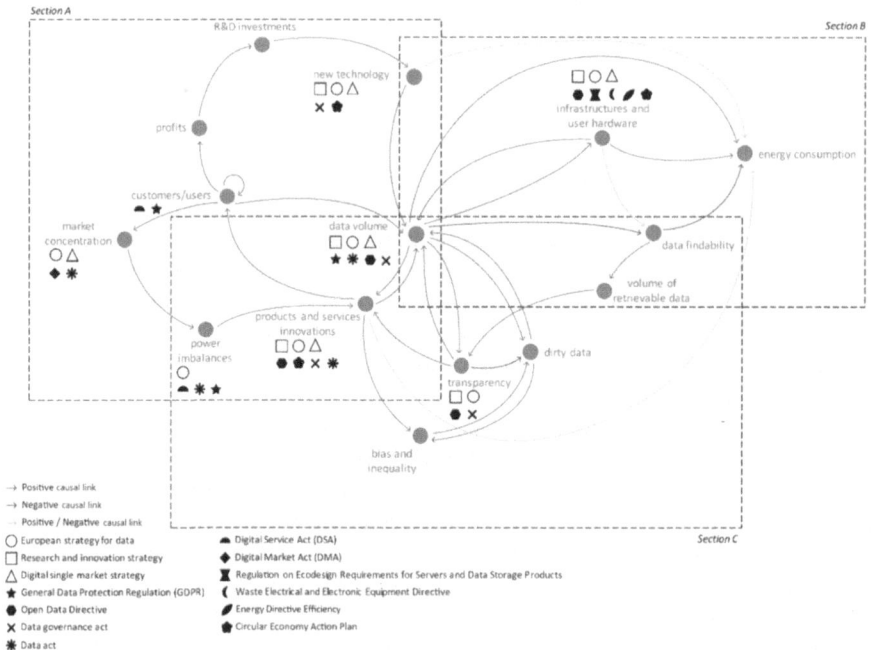

Fig. 1. Conceptual model.

4.1 The Data Economy

Section A schematises the mechanisms feeding the pace of the data economy that we know today: the increased volume of data allows businesses to offer more tailored products and services, enhancing the customer experience they can deliver and consequently attracting more customers. This means that the volume of data businesses can collect equally surges alongside their profits. Increased profits translate into expanding opportunities for investment in further technological development, thus fostering the continuous growth of data. New technology bringing about new data collection possibilities is easily illustrated with IoT devices, which grant corporations access to new kinds of customer data [34]. It is crucial to emphasise that the different model nodes aren't always associated with a single organisation but rather encompass the broader data economy ecosystem. It is not necessarily the organisation registering the customer increase and its consequent growth in customer data and profits that will directly boost research and development (R&D) investments that will result in new technology. However, since one organisation will have more monetary capacity, it will feed the whole sector by making more orders from its suppliers and generally increasing the market size, allowing its other actors to boost their activities. As an illustrative example, we use smartphones, the data collecting devices, running the Android OS, which is arguably what generates the network effects mainly thanks to its Play Store. In this case, the hardware and the software are complementary products produced by different companies operating in an ecosystem with different roles. Indeed, smartphone manufacturers make the highest R&D investments and provide the technology necessary to collect certain data types. In contrast, the operating system developer provides the platform which arguably plays the biggest role in client attraction, as revealed by the failure of the Microsoft phone, which, despite proposing high-quality, innovative designs, ran on its own OS and struggled to attract third-party apps [35]. Even though the volume of data is at the model's centre, any node can be considered as a starting and ending point of the feedback loop. For example, concerning public sector services, one could argue that the surge in the volume of data comes from the digitalisation of services, itself induced by political will, rather than the development of revolutionary technology. In this case, the feedback loop would start with the product and services innovation node. Alternatively, we could consider instances where the loop starts with increased customers. For example, such cases can happen with external factors influencing the market, as in several countries with governments subsidising electric vehicle purchases.

The extension of the left part of the section includes a highly simplified depiction of how the platform economy works. Indeed, the platform model has network externalities, which means that the perceived value offered by the platform increases as the number of users increases [36]. What characterises these platforms is that as the number of users grows, not only does the amount of data they can collect follows the same trend, but the platform will become increasingly attractive to other potential users and other market sides, which is a dynamic leading to market concentration and monopoly risks. Indeed, market concentration and the power it gives to large established companies allows them to acquire potential new market entrants, which could threaten their dominant position by eroding their user pool. This implies that companies that could change the status quo by bringing ethical solutions for users are often purchased by their competitors before they can grow enough to have an impact [37]. This concentration also results in power imbalances, with data subjects losing control over their data in favour of the data holders. This control loss can have consequences that data subjects would not expect and accept, such as transferring their data to third parties [38], prominently exemplified by the Cambridge Analytica scandal [11].

4.2 Data Storage

Section B mainly concerns data storage. A diversity of infrastructures and user hardware are required to store and share data, including data centres, mobile networks, user equipment, and data portals, to mention a few examples. Naturally, as the volume of data grows, so does the need to expand the data storage and transfer capacity. This implies that the volume of infrastructures and hardware manufacturing would increase, and in step, the amount of electronic waste. However, the features and governance of the infrastructures can also affect the volume of data. The promoters of data spaces, for example, which have recently received much attention, argue that pooling data across the actors of strategic industries can give birth to innovative initiatives which none of the participating parties could have enacted by itself. These new initiatives typically generate data [39]. Furthermore, the quality of the infrastructure influences the findability of the data it hosts, which we argue is likely to decrease with the volume of data stored. This is because of the concept of data overload, introduced above, according to which having access to the immense volume of data will make it harder for potential data users to find precisely what they were looking for, and scarce findability will lead users to resort to search queries that require energy use [40]. More generally, data storage activities have ecological consequences. One of the primary contributors to the data storage carbon footprint is the energy consumed by data centres. According to the IEA [27], data centres' energy use is going to follow a growing trend in the years to come despite high-efficiency improvements which have allowed data centres energy consumption to increase at a rate between 20% and 70% globally while the number of data centre workloads recorded an inflation of 340% over the same period. All activities related to data-related hardware and its use inherently have environmental effects. This is observable in the polluting processes linked to resource extraction, manufacturing, and data transportation, and the resources essential for their functioning, such as cooling systems that demand substantial energy consumption.

4.3 Social Implications

Our model's section C refers to the social implications related to the volume of data. As mentioned above, data are often the basis of innovations in the public and private sectors, as they can elaborate new business models, the digitalisation of services, or product customisation. In addition to yielding the economic benefits motivating data-based innovation, these novelties have (un)desirable social implications. For example, a desirable implication relates to the transparency node, an ideal pursued by many legislations encouraging public institutions to share their data openly [41]. Transparency and participatory governance, allowing citizens and other stakeholders to monitor and take part in government initiatives, is a frequent driver. But so is the social and commercial value of data. Indeed, governments also decide to open their data to capitalise on its commercial value, allowing stakeholders to reuse and innovate upon it. Our model does recognise the potential of open data and transparency to fuel innovation while also considering the possible undesirable implications, such as misuse of open data, which can lead to the creation of dirty data.

Other undesirable implications relate to the shift to the digital delivery of products and services, which has been documented to feed the digital divide, the gap between those with access to technology and digitalised products and services and those without [42]. Such a trend reinforces existing imbalances and isolation dynamics to the detriment of certain shares of the population, which are generally worse off than the others, as being poorer, part of minorities, or older, for example [42]. New analytical techniques have come with many challenges in recent years, magnifying exiting problems or introducing new ones. Automated algorithmic decision-making, today widely used in many different contexts, has been recognised to amplify further existing systematic issues [43, 44], while having the novel characteristics of being "unregulated and often hidden and invisible" ([45], p. 394), which makes it more difficult to obtain accountability for the problems it generates. The perverse effects of algorithm use are often attributed to the data on which the models were trained. Indeed, as algorithms use their training data to make decisions, they are likely to reproduce patterns encountered in the training dataset, which may be dirty, outdated, or reflect biases from the past [46, 47]. We argue that if dirty data feeds algorithmic harm, the opposite is true, as harmful algorithmic predictions will be stored in databases and feed the overall volume of data. The growing exposure to data and information can result in an overload having adverse effects on mental and physical health and productivity [48]. Moreover, the increasing quantity of dirty data also introduces risks of losing the ability to tell correct information from fake news and can hence lead to disinformation [48]. Lastly, the model considers that data-driven innovations can have different types of effects on the consumption of energy. Indeed, data can be used to innovate in many ways, and while some of these will enable efficiency gains, others will further increase energy needs.

4.4 Summing Up: What is Data Pollution?

While the nodes within the model interact to constitute the feedback loops, we need to recognise the interactions among the feedback loops themselves and their respective sections. Section A represents the mechanisms propelling the rapid expansion of the data economy, which directly influences the requisites for data storage and infrastructure, as addressed in section B. As the volume of data continues to increase, it results in a surge in the demand for data storage and infrastructure. This growth in data storage typically implies higher energy consumption, leading to adverse environmental consequences. In this context, the social implications elucidated in section C gain prominence. The social implications and its associated innovations are inextricably linked to the growth in data volume. As data volume increases, it serves to magnify the social challenges and disparities mentioned in section C, thereby reinforcing the interactive nodes at play within the conceptual model. We hence come to the following synthetic definition:

> Data pollution is the set of harms generated by economic activities related to data collection, storage and use, which transversally impacts individuals and their living environments.

5 Current Policy Responses to Data Pollution

Overall, the complexity of data pollution arises from the interactions between its components, represented as nodes in our model, which are not easily separable given their circularity formed through feedback loops. This is especially true given that an effect can feed back into its cause. Because of this complexity, we believe it is necessary to tackle it comprehensively. However, to do so, there is a need to understand what is currently in place in the EU to address the complex mechanisms generating data pollution.

Firstly, and considering the rapid expansion of the data-driven economy, the EU has adopted a Data Strategy in 2020 aiming to foster data economy by creating a single data market, enhancing Europe's competitiveness and data sovereignty [49, 50]. The Data Governance Act and the Data Act serve as a legal framework to enable the practical implementation of the EU Data Strategy [51, 52]. The Data Governance Act aims to reinforce the single market for data notably by setting the conditions for common European data spaces aiming to facilitate the exchange of public sector and business data across the EU to stimulate the growth of novel data-driven products and services [5, 6]. Indeed, as the volume of data generated from digital devices and services continues to grow, there was a need for a legal framework aiming to harness the potential of data to the benefit of the EU economy and society, and to avoid reliance on third countries, particularly in the advancement of IoT or AI systems [53]. In this aspect, the Data Act complements the Data Governance Act by setting rules for using data generated by IoT to boost the EU's data economy [54]. The Data Act also fits the idea of growing the EU data-driven economy by creating avenues and eliminating obstacles for data reuse [55]. Complementarily for the public sector and publicly funded data, the Open Data Directive completes the desire of the EU to foster data sharing and data-driven innovation across all sectors of the economy by promoting the availability and reusability of public sector data in the EU [56]. The Open Data Directive is a central instrument for the realisation of the EU data economy by underlying the willingness of the EU to capitalise on public sector data to feed the data economy [57]. By doing so, it also gives the right to individuals to access information retained by public authorities, facilitating government transparency and accountability, nurturing public trust, and, in turn, expanding public engagement [58]. While these documents do not explicitly address data pollution, they still contain elements and objectives that may indirectly contribute to mitigating the dark sides of the data-driven economy and the data pollution it generates. For example, they encourage responsible data sharing through European data spaces, which may reduce data fragmentation and inefficiency, contributing to data pollution. They also encourage responsible data management practices such as the publication of high-quality data, which indirectly reduce the risk of data pollution arising from the dissemination of dirty data and reduce the risks of misuse, such as the creation of bias or inaccurate analysis contributing to misinformation. Through their broader goals of promoting responsible data management practices, these documents can indirectly contribute to reducing data pollution. There is however no specific provision addressing data pollution as a standalone issue.

Secondly, as the volume of data increases, the demand for infrastructure and user hardware also grows in proportion to the data volume. However, data storage activities generate data pollution by having ecological consequences. The EU only addresses indirectly data pollution generated by data storage only through broader environmental and sustainable efforts. For example, the Circular Economy Action Plan encourages responsible product design and management, including electronic and IT equipment [59]. Other examples are the Regulation on Ecodesign Requirements for Servers and Data Storage Products, which aims to limit the environmental impacts of servers and data storage products by setting rules on energy efficiency [60] or the revised Energy Efficiency Directive, which tackles the heating and cooling of data centres by introducing an obligation for the monitoring of the energy performance of data centres to ensure a fully decarbonised heating and cooling supply by 2050 [61]. These examples show how environmental aspects of data pollution related to data storage are typically addressed through broader regulations setting energy efficiency targets and emissions reduction goals or tackling distinct problems such as responsible data management (e.g., GDPR, Digital Service Act (DSA), Digital Market Act (DMA)) or the recycling of electronic equipment used for data storage through the Waste Electrical and Electronic Equipment Directive [62–65].

Thirdly, the quest to foster a data-driven economy has created social implications by generating an environment in which some data giants have unique control over some data, which is no longer offset by the control of other actors [50]. As technology has become increasingly complex and invasive, especially given that various businesses exploit data through algorithmic decision-making, it is increasingly complicated for consumers and users to maintain control over their data [50]. This led to the emergence of a darker narrative around the data-driven economy, especially as individuals encounter daily vast amounts of information across various devices and media, which can overwhelm them and thus jeopardise their ability and motivation to scrutinise essential details for informed decisions and instead chose defaults' options which are presented to them as recommendations [50]. This leads individuals to often provide consent without considering the consequences, especially when faced with consent requests. Considering that data are nonrival and the potential benefit they hold for the economy, personal data are widely shared, reused and hence risk being misused. To address these concerns, the EU introduced the GDPR in 2018, aiming to regulate the processing of personal data within the EU, granting consumers enhanced protection and control over their data [63]. The GDPR sets standards for data privacy by covering any organisation that collects or processes EU citizens' data independently of the organisation's location [66]. The GDPR primarily focuses on protecting individuals' data, ensuring their privacy rights are respected and providing mechanisms for individuals to control how their data are collected, processed, and used [66]. Other regulations have entered into force to strengthen online rights and regulate digital services, such as the DSA and the DMA [64, 65]. The DSA aims to regulate digital platforms to protect individuals and their fundamental rights online while fostering innovation, growth, and competitiveness. The DMA is another legislative framework targeting digital platforms to ensure fair competition. While these legislative frameworks primarily focus on protecting rights and regulating digital services rather than addressing data pollution as a separate issue, they include some provisions that may indirectly help address some aspects of data pollution. For example, the GDPR promotes the right to erasure, which ensures that personal data can be erased under some circumstances, such as if the personal data are no longer necessary for the purposes for which they were collected. Another illustrative example is the principle of data minimisation, which encourages organisations to collect and process only the data that is directly relevant and necessary to accomplish a specified purpose. Hence, by intending to limit excessive data collection, the GDPR may reduce the data pollution generated by accumulating unnecessary data. As for the DSA, regulating digital platforms, notably through content regulation and platform accountability, it indirectly addresses data pollution through provisions aiming at reducing harmful or misleading data, leading to the spread of illegal content and disinformation. Finally, while the primary objective of the DMA is to address competition issues in the digital market, it also indirectly addresses data pollution by fostering responsible data management and ensuring fair access to data.

6 Conclusion

Data pollution is a term that has been used since the 1980s in different domains and contexts and has hence been given different meanings over time. As data pollution has always been considered within some specific contexts of interest of the authors who wrote about it, but never in its globality, it has only enabled the conceptualisation of partial solutions to the set of harms entailed in data pollution. However, while being different, the meanings attributed to data pollution have a similarity: they describe harms produced as byproducts of the data economy. With this article, we propose a conceptualisation of data pollution, positioning it within the context from which it emerges and that perpetuates, it and contending that all the different harms produced by the data activities not only originate from interdependent mechanisms but also result from and feed one another. As a result of this conceptualisation, we define data pollution as the set of harms originating from any data activity. We believe such a holistic approach is necessary for policymakers to conceive ways to address data pollution effectively.

We argue that the current EU regulation primarily focuses on specific aspects of data pollution while ignoring other large strands of the phenomenon and is not fit to grant adequate protection against many of the harms of data pollution. Although concerns over data subjects' privacy and considerations over the competitiveness of infrastructures and European companies are addressed by the law, ecological harms caused by data activities and many social adverse effects, including data exclusion and information overload, are not. Indeed, while the various law texts each address specific aspects of data pollution, their provisions are not directly related and do not account for the interrelated nature of the harms stemming from data pollution. To ensure that the continuous digitalisation of our activities allows the transition to a more ecological and socially fair society, we it is essential to understand data pollution comprehensively and address it as such, with a coordinated regulatory approach. We hope that the concept of data pollution as presented in this article can serve as the basis for future research investigating the possibilities to create regulation addressing data pollution comprehensively.

Acknowledgments. This study was funded by the Swiss National Science Foundation (grant number 212637).

Disclosure of Interests. The authors have no competing interests to declare that are relevant to the content of this article.

References

1. Tan, K.H., Ji, G., Lim, C.P., Tseng, M.L.: Using big data to make better decisions in the digital economy. Int. J. Prod. Res. **55**(17), 4998–5000 (2017)
2. Calzada, I., Almirall, E.: Data ecosystems for protecting European citizens' digital rights. Transforming Gov.: People Process Policy **14**(2), 133–147 (2020)
3. Curry, E., et al.: The European big data value ecosystems. In: Curry, E., Metzger, A., Zillner, S., Pazzaglia, J.C., Robles, A.G. (eds.) The Elements of Big Data Value, pp. 3–20 (2021)
4. The Economist Group: The future of Europe's data economy (2022). https://impact.economist.com/perspectives/sites/default/files/ei233_msft_futuredata_report_-_v7.pdf. Accessed 08 Nov 2023
5. European Commission: Building a data economy - Brochure (2019). https://digital-strategy.ec.europa.eu/en/library/building-data-economy-brochure. Accessed 08 Nov 2023
6. Common European data spaces. https://dataspaces.info/common-european-data-spaces/#page-content. Accessed 08 Nov 2023
7. Andersen, K.N., et al.: Fads and facts of e-government: a review of impacts of e-government (2003–2009). Int. J. Public Adm. **33**(11), 564–579 (2010)
8. Baku, A.A.: Digitalisation and new public management in Africa. In: Hinson, R.E., Madichie, N., Adeola, O., Nyigmah Bawole, J., Adisa, I., Asamoah, K. (eds.) New Public Management in Africa, pp. 299–316. Springer, Cham (2022). https://doi.org/10.1007/978-3-030-77181-2_12
9. Global Open Data Index. http://index.okfn.org/place.html. Accessed 04 Mar 2024
10. ThoughtLab: Building a Future-Ready City (2022). https://thoughtlabgroup.com/wp-content/uploads/2022/11/Building-a-Future-Ready-City-ebook_Final-November-2022.pdf. Accessed 04 Mar 2024
11. Ben-Shahar, O.: Data pollution. J. Legal Anal. **11**, 104–159 (2019)
12. Hasselbalch, G.: Data Pollution & Power: White paper for a global sustainable agenda on AI (2022). https://gryhasselbalch.com/books/data-pollution-power-a-white-paper-for-a-global-sustainable-development-agenda-on-ai/. Accessed 08 Nov 2023
13. The Shift Project: Lean ICT - Towards Digital Sobriety (2019). https://theshiftproject.org/wp-content/uploads/2019/03/Lean-ICT-Re%ADport_The-Shift-Project_2019.pdf
14. Amoroso, D.L., Mcfadden, F., White, K.B.: Disturbing realities concerning data policies in organizations. Inf. Resour. Manag. J. (IRMJ) **3**(2), 18–28 (1990)
15. Zimmerli, W.C.: Who is to blame for data pollution? On individual moral responsibility with information technology. In: Mitcham, C., Huning, A. (eds.) Philosophy and Technology II, vol. 90, pp. 291–305. Springer, Dordrecht (1986). https://doi.org/10.1007/978-94-009-4512-8_21
16. Cao, Y., et al.: Efficient repair of machine learning systems via causal unlearning. In: Kim, J., et al. (eds.) Asia Conference on Computer and Communications Security, Incheon, Republic of Korea, pp. 735–747. Association for Computing Machinery (2018)
17. Esfahani, A., Mantas, G., Rodriguez, J., Nascimento, A., Neves, J.C.: A null space-based MAC scheme against pollution attacks to random linear network coding. In: Swantee, O., Wang, J. (eds.) International Conference on Communication Workshop, London, UK, pp. 1521–1526. IEEE (2015)
18. Chang, J., et al.: Application of deep machine learning for the radiographic diagnosis of periodontitis. Clin. Oral Invest. **26**(11), 6629–6637 (2022)
19. De Nadai, A.S., Hu, Y., Thompson, W.K.: Data pollution in neuropsychiatry—an under-recognized but critical barrier to research progress. JAMA Psychiat. **79**(2), 97–98 (2022)
20. Lin, H., Ai, Y., Ling, Z.: A light CNN with split batch normalization for spoofed speech detection using data augmentation. In: Theera-Umpon, N., et al. (eds.) Asia-Pacific Signal and

Information Processing Association Annual Summit and Conference, Chiang Mai, Thailand, pp. 1684–1689. IEEE (2022)
21. Dong, J., Curtmola, R., Nita-Rotaru, C., Yau, D.K.Y.: Pollution attacks and defenses in wireless interflow network coding systems. Trans. Dependable Secure Comput. **9**(5), 741–755 (2012)
22. Castelluccia, C., Kaafar, M.A.: Owner-centric networking (ONC): toward a data pollution-free internet. In: Yoshida, K., Chang, M. (eds.) Ninth Annual International Symposium on Applications and the Internet, Bellevue, USA, pp. 169-172. IEEE (2009)
23. Shenk, D.: Data Smog: Surviving the Information Glut. Harper, San Francisco (1998)
24. Gouws, R.H., Tarp, S.: Information overload and data overload in lexicography. Int. J. Lexicogr. **30**(4), 389–415 (2017)
25. Bawden, D., Holtham, C., Courtney, N.: Perspectives on information overload. ASLIB Proc. **51**, 249–255 (1999)
26. "Climate crisis: The unsustainble use of online video": Our new report on the environmental impact of ICT. https://theshiftproject.org/en/article/unsustainable-use-online-video/. Accessed 24 Oct 2023
27. Data centres and data transmission networks. https://www.iea.org/energy-system/buildings/data-centres-and-data-transmission-networks. Accessed 24 Oct 2023
28. Bietti, E., Vatanparast, R.: Data waste. Harvard Int. Law J. Front. **61**, 1–11 (2020)
29. François, C.: Systemics and cybernetics in a historical perspective. Syst. Res. Behav. Sci.: Official J. Int. Federation Syst. Res. **16**(3), 203–219 (1999)
30. Heylighen, F., Joslyn, C.: Cybernetics and second-order cybernetics. In: Encyclopedia of Physical Science & Technology, pp. 155–170 (2001)
31. Mitchell, M.: Complex systems: network thinking. Artif. Intell. **170**(18), 1194–1212 (2006)
32. Glanville, R.: Second order cybernetics. Syst. Sci. Cybern. **3**, 59–85 (2002)
33. Scott, B.: Second-order cybernetics: an historical introduction. Kybernetes **33**(9/10), 1265–1378 (2004)
34. Rose, K., Eldridge, S., Chapin, L.: The internet of things: an overview (2015). https://www.internetsociety.org/wp-content/uploads/2017/08/ISOC-IoT-Overview-20151221-en.pdf. Accessed 08 Nov 2023
35. Windows phone was a glorious failure: Looking back on the bumpy road taken by Microsoft's most ambitious mobile OS. https://www.theverge.com/2017/10/10/16452162/windows-phone-history-glorious-failure. Accessed 05 Nov 2023
36. Rietveld, J., Schilling, M.A.: Platform competition: a systematic and interdisciplinary review of the literature. J. Manag. **47**(6), 1528–1563 (2021)
37. Katz, M.L.: Multisided platforms, big data, and a little antitrust policy. Rev. Ind. Organ. **54**(4), 695–716 (2019)
38. Papadopoulou, E., Stobart, A., Taylor, N.K., Williams, M.H.: Enabling data subjects to remain data owners. In: Jezic, G., Howlett, R., Jain, L. (eds.) Agent and Multi-agent Systems: Technologies and Applications, vol. 38, pp. 239–248. Springer, Cham (2015). https://doi.org/10.1007/978-3-319-19728-9_20
39. European Commission, Communication from the Commission to the European Parliament, the Council, the European Economic and Social Committee and the Committee of the Regions: "Towards a common European data space", COM(2018) 232 final, Brussels, Belgium, pp. 1–15 (2018)
40. Google: Environment Report (2023). https://www.gstatic.com/gumdrop/sustainability/google-2023-environmental-report.pdf. Accessed 26 Oct 2023
41. Attard, J., Orlandi, S., Scerri, S., Auer, S.: A systematic review of open government data initiatives. Gov. Inf. Q. **32**(4), 399–418 (2015)
42. Cullen, R.: Addressing the digital divide. Online Inf. Rev. **25**(5), 311–320 (2001)

43. Donovan, J., Caplan, R., Matthews, J., Hanson, L.: Algorithmic accountability: a primer (2018). https://datasociety.net/wp-content/uploads/2019/09/DandS_Algorithmic_Accountability.pdf. Accessed 08 Nov 2023
44. Sepúlveda Carmona, M., et al.: Extreme poverty and human rights: Note by the Secretary-General (2011). https://digitallibrary.un.org/record/1648309?ln=en. Accessed 08 Nov 2023
45. Marjanovic, O., Cecez-Kecmanovic, D., Vidgen, R.: Algorithmic pollution: making the invisible visible. J. Inf. Technol. **36**(4), 391–408 (2021)
46. The police are using computer algorithms to tell if you're a threat. https://time.com/4966125/police-departments-algorithms-chicago/. Accessed 10 Nov 2023
47. Rosenblat, A., Wikelius, K., Boyd, D., Gangadharan, S.P., Yu, C.: Data & civil rights: criminal justice primer. In: Boyd, D., Peña Gangadharan, S., Yu, C. (eds.) Data & Civil Rights Conference, Washington, DC (2014)
48. Bawden, D., Robinson, L.: Information overload: an overview. In: Oxford Encyclopedia of Political Decision Making. Oxford University Press, Oxford (2020)
49. European data strategy. https://commission.europa.eu/strategy-and-policy/priorities-2019-2024/europe-fit-digital-age/european-data-strategy_en#data-governance. Accessed 26 Oct 2023
50. Van Ooijen, I., Vrabec, H.U.: Does the GDPR enhance consumers' control over personal data? An analysis from a behavioural perspective. J. Consum. Policy **42**, 91–107 (2019)
51. European Commission, Proposal for a Regulation of the European Parliament and of the Council on harmonised rules on fair access to and use of data (Data Act) (2022)
52. Official Journal of the European Union, Regulation (EU) 2022/868 if the European Parliament and of the Council of 30 May 2022 on European data governance and amending Regulation (EU) 018/1724 (Data Governance Act), in: Official Journal of the European Union (Ed.), pp. 1–44 (2022)
53. Baloup, J., et al.: White paper on the data governance act (2021). https://papers.ssrn.com/sol3/papers.cfm?abstract_id=3872703. Accessed 08 Nov 2023
54. European Commission, Data Act: Commission welcomes political agreement on rules for a fair and innovative data economy, European Commission, Brussels (2023)
55. The European Data Act. https://www.eu-data-act.com/. Accessed 07 Nov 2023
56. van Eechoud, M.: A serpent eating its tail: the database directive meets the open data directive. IIC-Int. Rev. Intellectual Property Competition Law **52**(4), 375–378 (2021)
57. Official Journal of the European Union, Directive (EU) 2019/1024 of the European Parliament and of the Council of 20 June 2019 on open data and the re-use of public sector information, in: Official Journal of the European Union (Ed.), pp. 1–28 (2019)
58. Broomfield, H.: Where is open data in the Open Data Directive? Inf. Polity **28**(2), 175–188 (2023)
59. European Commission, Communication from the Commission to the European Parliament, the Council, the European Economic and Social Committee and the Committee of the Regions: A new Circular Economy Action Plan - For a cleaner and more competitive Europe, Brussels (2020)
60. Official Journal of the European Union, Commission Regulation (EU) 2019/424 of 15 March 2019 laying down ecodesign requirements for servers and data storage products pursuant to Directive 2009/125/EC of the European Parliament and of the Council and amending Commission Regulation (EU) No 617/2013, in: Official Journal of the European Union (Ed.), pp. 1–21 (2019)
61. Energy efficiency directive. https://energy.ec.europa.eu/topics/energy-efficiency/energy-efficiency-targets-directive-and-rules/energy-efficiency-directive_en#heating-and-cooling-and-data-centres. Accessed 07 Nov 2023

62. Official Journal of the European Union, Directive 2012/19/EU of the European Parliament and of the Council of 4 July 2012 on waste electrical and electronic equipment (WEEE), in: Official Journal of the European Union (Ed.), pp. 1–34 (2012)
63. Official Journal of the European Union, Regulation (EU) 2016/679 of the European Parliament and the Council of 27 April 2016 on the protection of natural persons with regard to the processing of personal data and on the free movement of such data, and repealing Directive 95/46/EC (General Data Protection Regulation), pp. 1–88 (2016)
64. Official Journal of the European Union, Regulation (EU) 2022/2065 of the European Parliament and of the Council of 19 October 2022 on a Single Market For Digital Services and amending Directive 2000/31/EC (Digital Services Act), in: Official Journal of the European Union (Ed.), pp. 1–102 (2022)
65. Official Journal of the European Union, Regulation (EU) 2022/1925 of the European Parliament and of the Council of 14 September 2022 on contestable and fair markets in the digital sector and amending Directives (EU) 2019/1937 and (EU) 2020/1828 (Digital Markets Act), in: Official Journal of the European Union (Ed.), pp. 1–66 (2022)
66. Zaeem, R.N., Barber, K.S.: The effect of the GDPR on privacy policies: recent progress and future promise. ACM Trans. Manag. Inf. Syst. (TMIS) **12**(1), 1–20 (2020)

Understanding the Problem Space for Effective Use of a Circular Economy Monitor in Policy Making

Michiel Pauwels[1,3](✉), René Reich[2,4], An Vercalsteren[3], Maarten Christis[3], Luc Alaerts[2,4], and Karel van Acker[1,2,4]

[1] Center for Economics and Corporate Sustainability, Faculty of Economics and Business, KU Leuven, Leuven, Belgium
michiel.pauwels@kuleuven.be
[2] Sustainability Assessments of Materials and Circular Economy, Department of Materials Engineering, KU Leuven, Leuven, Belgium
[3] Unit of Sustainable Materials and Chemistry, VITO, Mol, Belgium
michiel.pauwels@vito.be
[4] Flanders Make Vzw, VCCM Corelab, Leuven, Belgium

Abstract. This paper identifies and validates the challenges hindering the integration of circular economy into evidence-based policy making, and proposes an outlook for enhancing the effective use of circular economy monitors. It highlights the limitations of current circular economy monitoring systems, which often fail to transform circular economy information into actionable knowledge for policy makers. Using the echelon Design Science Research approach, which divides projects into manageable 'echelons' to tackle complex socio-technical problems, this study focuses on the problem analysis echelon. Through 13 semi-structured interviews with intended users and an extensive literature review, the study identifies and validates five challenges to embedding circular economy in policymaking. These challenges are the delayed benefits of circular economy actions, fragmented policy coordination, the lack of a policy agenda for higher R strategies, the complexity of circular economy implementation, and the gap between theoretical frameworks and practical policy needs. This analysis is grounded in the theory of effective use as our Kernel Theory. To address the ineffective use of circular economy monitors, the study proposes an outlook of design requirements and design principles. This paper contributes to the literature on policy monitoring frameworks and circular economy policymaking by delineating a validated problem space that future researchers can use to improve the effective use of circular economy monitors in policymaking.

Keywords: Circular Economy · Decision Support System · Design Science Research

1 Introduction

The transition to a resource-efficient and climate-neutral society represents a systemic change affecting all aspects of society. The circular economy (CE) promises an alternative economic system in which resources are kept at the highest possible level of functionality at every part in the value chain, with the aim of achieving environmental sustainability and socioeconomic prosperity [1]. While the CE concept has become popular with policymakers and industry alike, the implementation of a CE requires fundamental changes in the way current value chains and network's function. Despite general agreement that public policy plays a key role in facilitating CE transition, one of the main challenges remains the development and implementation of policies and regulatory instruments that contributes to the CE transition [2]. The effective integration of CE principles into policy processes requires data-driven decisions grounded in measurable criteria. Therefore, a robust policy monitoring system, equipped with indicators, is indispensable to furnish policymakers in making data-driven CE decisions [3]. Through such a system, governments are provided with the material properties of the CE and the functional affordance for improved decision making [4]. Although monitors summarize evidence about the state of the CE, policymakers find it difficult to draw conclusions from data [5] and identify measures to intervene [6]. This highlights a gap in the literature regarding the practical use of monitoring systems and their influence on policy decisions. This paper presents an ongoing policy research collaboration dedicated to the development of a CE monitoring system for Flanders, aimed at providing direct feedback on the CE to policymakers [7]. To manage the complexity of our project, we used the echeloned DSR (eDSR) approach [8], focusing on the problem analysis echelon, to develop a validated problem statement. Our starting point is the online prototype of the CE monitor (https://cemonitor.be/). We conducted 13 semi-structured interviews with intended users of the monitor to evaluate its usage and identify potential use cases. This analysis is grounded in Burton-Jones and Grange's [9] Theory of Effective Use (TEU) as Kernel Theory. The findings from these interviews are validated in the literature and abstracted into a broader problem statement, addressing the issues underlying the integration of CE into evidence-based policymaking. Following this, we propose an outlook for Design Requirements (DRs) and Design Principles (DPs) to improve the effective use of CE monitors.

2 Theoretical Background

2.1 CE Policy Making

In response to the global necessity of managing natural resources more efficiently and transitioning to climate-neutral societies, numerous governments have declared the political ambition to transition towards a CE. The EU, in particular, has placed the CE high on its political agenda since 2015, following the introduction of the Circular Economy Action Plan [10]. It is widely agreed that governmental interventions have an important role to play in facilitating the CE [11, 12]. Despite general agreement that public policy plays a key role in facilitating CE transition, one of the main challenges remains the development and implementation of policies and regulatory instruments that influences material resource efficiency [2]. The effectiveness of government intervention and the transition progress require data-driven policy decisions and monitoring to ensure continuous progress. Hence, public CE governance implies the challenge of assessing this new form economy.

2.2 Data-Driven Policy Making

Governments are increasingly acknowledging the potential of data utilization in enhancing policymaking and citizen service delivery [13]. Politicians use data to make national and international comparisons, reason arguments, and draft new initiatives. Policymakers use them to design, implement, and assess legislation. On the other hand, citizens can judge whether the policies and the responsible party serve them well based on indicators [14]. Data-driven policymaking, emphasizes the use of data analytics, big data, and open data across the three successive stages of policymaking: agenda setting, policy implementation, and policy evaluation [15]. Data-driven policies can boost transparency, legitimacy, and efficiency. More importantly, the transition to a CE necessitates appropriate, environmentally evidence-based policies and strategies for achieving a more sustainable society [13]. Data pertaining to the environment, infrastructure, technology, economy, and citizen behavior are crucial in building and reinforcing the local government's knowledge base [16, 17]. Data-driven policymaking builds on the concept of evidence-based policymaking, which considers three types of evidence: *"systematic ('scientific') research, program management experience ('practice'), and political judgement"* [15, p. 1]. Consequently, it also faces several challenges of evidence-based policymaking extended by data-related issues, such as:

- a shift in culture and process of policymaking [19],
- social inclusion, involvement, and co-creation by citizens and stakeholders [15],
- the growing number of necessary data sources and involved stakeholders with increasing policy complexity [20, 21],
- lacking skills of policymakers to draw conclusions from data [5].
- the unclear conversion of qualitative and quantitative data into policies [20],

Data-driven policymaking is therefore a prerequisite for CE policy making. To this end, it is necessary to develop indicators and monitor progress in order to implement CE policies effectively [3].

2.3 CE Monitoring Tools for Governments

Monitors or indicator dashboards collect, aggregate and deliver evidence about the CE to the policymakers [1]. By tracing the results' developments of different indicators over time and allowing for comparisons to other governments, monitors contribute to the outcome assessment, action formation, and belief formation of the CE transition [4]. The first instances of CE monitoring systems were introduced in Japan in 2000 [22], followed by China in 2008 [23]. This practice was subsequently adopted by various European Union (EU) member states, including France [24], the Netherlands [25], Flanders [7], and the EU itself through its monitoring framework [26]. These monitoring frameworks vary in format, ranging from periodically updated, publicly accessible PDFs [24, 25] to interactive online portals [7, 26]. Although monitors summarize evidence about the state of the CE, policymakers find it difficult to draw conclusions from data [5] and identify measures to intervene [6]. The underlying assumption is that once a CE monitoring system is published, it will support the CE transition by informing policy departments. However, the validity of this assumption remains uncertain as there is a lack of supporting evidence in the literature on how monitoring systems are actually used in practice and whether they influence policy decisions. We address this gap using a DSR approach to evaluate the use of the CE monitor in Flanders and propose an outlook for DRs and DPs to improve the effective use of the monitor.

2.4 Theory of Effective Use

Information systems (IS) don't exist for their own sake but are always used to achieve a higher goal [9]. Burton-Jones & Gregor [9] developed a theory to test and improve the effective use of an IS. They define effective use as "using a system in a way that helps attain the goals for using the system" [9, p. 633]. Their framework encompasses three dimensions that influence the effective and efficient use of IS: (1) transparent interaction, (2) representational fidelity, and (3) informed action. Transparent interaction refers to the user's ability to access the system's representation through its surface and physical structures. Representational fidelity pertains to the user's ability to obtain accurate information about the represented domain from the system. Informed action involves the user's ability to translate the representation into actions that improve a particular state. Burton-Jones and Grange [9] identify two approaches to improve the effective use of an IS: adaptation and learning. In adaptation the user changes the representation of the relevant domain or the access to the representation. Learning consists of actions taken by the users to understand the 1) system, 2) the domain it represents 3) the fidelity of its representation, and 4) how to use the representation [9].

3 Research Design

Our research is guided by the Design Science Research (DSR) paradigm. DSR aims to develop design knowledge that informs the creation of IS artifacts such as software, methods, models, or concepts [27]. The iterative nature of DSR allows the researcher to evaluate and improve the artefact throughout the DSR project. Our DSR project is an ongoing policy research collaboration dedicated to developing a CE monitoring system for Flanders, aimed at providing direct feedback on CE to policymakers [7]. The initial design cycle focused on developing the CE monitoring concept and implementing version 1.0 of the online prototype, as detailed in Alaerts et al. [6]. This paper initiates a new design cycle to improve the online prototype and create version 2.0 of the monitor. We employed the recently developed eDSR methodology [8], which divides DSR projects into smaller "echelons" to facilitate the development of impactful artifacts addressing complex socio-technical challenges. The eDSR methodology breaks down the DSR projects in five echelons (1) problem analysis, (2) definition of objectives and requirements, (3) design & development, (4) demonstration, and (5) evaluation. This paper focuses on the problem analysis echelon as understanding the problem of how CE monitors are used in policy making are essential for improving their effective use in evidence-based policy making. We use the online available CE monitor (https://cemonitor.be/) of Flanders as a case study to evaluate the effective use of the monitor. For the problem analysis, we conducted between 25 March and 29 April thirteen semi-structured interviews with intended users of the monitor. The purpose of these interviews was to assess the monitor's impact on the policy-making process. The interview guideline focused on three main themes: policy processes for the CE, information needs for these policy processes, and the use and perception of the CE Monitor. For the purposes of this paper, these interviews were selectively open coded, focusing on the user's experience and perception of the monitor (4.1). To identify potential effective use cases of the CE monitor in policymaking (4.2), our research adopted the TEU framework [9]. Findings from the interviews are then enriched and validated with a targeted literature review. Our literature review focused on the disturbances of integrating CE into governmental policymaking. In TEU, disturbances are uncontrollable or unpredictable elements that hinder the effective use of an IS [9]. We used the Scopus database for this literature search, employing the search terms "circular economy" AND "policy mak*" AND "government" OR "public administration" AND "challenges" OR "barriers" OR "obstacles". This search yielded 111 articles, from which we selected 15 based on their abstracts. This resulted in problem identification (4.3) and a validated problem statement that marks the end of the problem analysis echelon. To enhance the effective use of the CE monitor, we propose an initial outlook on the objectives and requirements echelon. Drawing from the TEU framework as the kernel theory, we derived a set of Design Requirements (DRs) (5.1) and formulated Design Principles (DPs) (5.2) to improve the effective use of CE monitors.

4 Problem Analysis

In the upcoming section, we will address the issues identified from our interviews with policymakers on evaluating the use of a CE monitor, as well as insights obtained from existing literature on embedding CE in evidence-based policy making.

4.1 Understanding and Defining the Policymaker's Problem

Our collaboration with Flemish policymakers was initiated in 2017 in light of a mutual interest in embedding CE in the Flemish policy domain. The interaction between research and policy was driven by the ambition of policy makers to measure the progress of Flanders in the transition to a CE. As a consequence, the main purpose for the instantiation of the monitor was to provide more direct policy feedback on the emerging and rapidly evolving topic of the CE in Flanders and to provide information for evidence-based policymaking [7]. Drawing on Weber's representation theory which states that the purpose of an IS is to faithfully represent a real-world domain [28], the main purpose of the CE monitor is to faithfully represent the CE domain in Flanders. According to Weber's theory [28], an IS comprises three structures: 1) the deep structure, which specifies the domain represented by the IS, 2) the surface structure, which includes the features that enable user access and interaction with the representation, and 3) the physical structure, which encompasses the hardware supporting the other structures. The deep and surface structures of the monitor were developed through various iterative design cycles involving policymakers and academics [6]. The deep structure of the monitor includes approximately 100 indicators, organized into a top layer with the themes 'circularity' and 'effects,' along with four Need Satisfier Systems (NSS) that contain more detailed indicator layers. Each layer is divided into sections with one to ten indicators. A dedicated panel contextualizes and explains the data, offering technical details and references for each indicator. In terms of the surface structure, the monitor was publicly launched in 2021 and is accessible through the Monitor's online portal (https://cemonitor.be). To this end, the portal allows users to access and interact with the information represented on the monitor. Although the monitor is available to the broader public, the target audience is primarily Flemish policy makers. To derive benefits from the monitor, it must be used effectively [9]. Hence, the deep structure of the monitor provides information on the CE by integrating data and relevant context about the CE. Policymakers should thus be able to access this information through the monitor's surface structure and translate it into knowledge for evidence-based policy decisions. However, interviews with 13 intended users of the monitor revealed that in reality the monitor was limitedly consulted. Only 5 out of the 13 interviewees reported having accessed the monitor during the recent period. As one interviewee stated: *"I recall visiting it initially a couple of years ago at its first publication. A few months ago, I checked it again for any new updates, but aside from that, I don't use it regularly"*. Moreover, almost none of the interviewees could cite an example of the CE monitor directly influencing a policy decision. Most interviewees acknowledged that, while they did not use the monitor themselves, they assumed it was utilized by other policymakers. One interviewee noted, *"I must admit that I barely use it, honestly. But I guess other policymakers do use the monitor more because their tasks are much more policy-related, and they need that kind of information."* What was also striking was that although most interviewees admitted that they did not use the monitor regularly, almost all of them stressed the necessity for having such a monitoring system for policy making. As one interviewee noted, *"We are interested in the monitor, but there is currently no systematic processing of reports and information from the monitor."* Hence, these statements illustrate that, despite the availability of the CE monitor, the information representation of the CE in Flanders remains underutilized. This ineffective use of the monitor hampers the efforts to integrate CE information into evidence-based decision-making. Therefore, we define the following problem situation:

Definition of the Problem Situation: The lack of effective use of the CE monitor in policy making decisions. Addressing this problem first requires a better understanding of the potential reasons why policymakers intend to use a monitoring system. Once these reasons are clear, we abstract from the specific case of the monitor as an instrument for effective decision-making and look in the literature for issues and challenges that occur in embedding CE in effective policy making processes.

4.2 Identifying the Potential Effective Use Cases of a CE Monitor

Burton-Jones and Grange [9] define effective use and performance as objectives of creating an IS. Effective use is seen as the lower-level goal that serves as a means, while performance is the higher-level desired goal. Alaerts et al. [7] describe that the main objective of developing the CE monitor was to provide more direct policy feedback on CE and provide evidence-based information for policymaking, thereby improving policy maker's performance in the CE domain. Policymakers accessing these monitoring systems typically seek to satisfy specific information needs, reflecting potential effective use cases of the monitor. As illustrated in Sect. 4.1, the ineffective use of the CE monitor hinders the achievement of the higher goal of embedding CE in policy making. Figure 1 illustrates the adaptation of Burton-Jones and Grange's [9] hierarchical dimensional framework for studying effective use and its impact on performance to the context of the CE monitor. The potential effective use cases of the monitor were identified during the interviews.

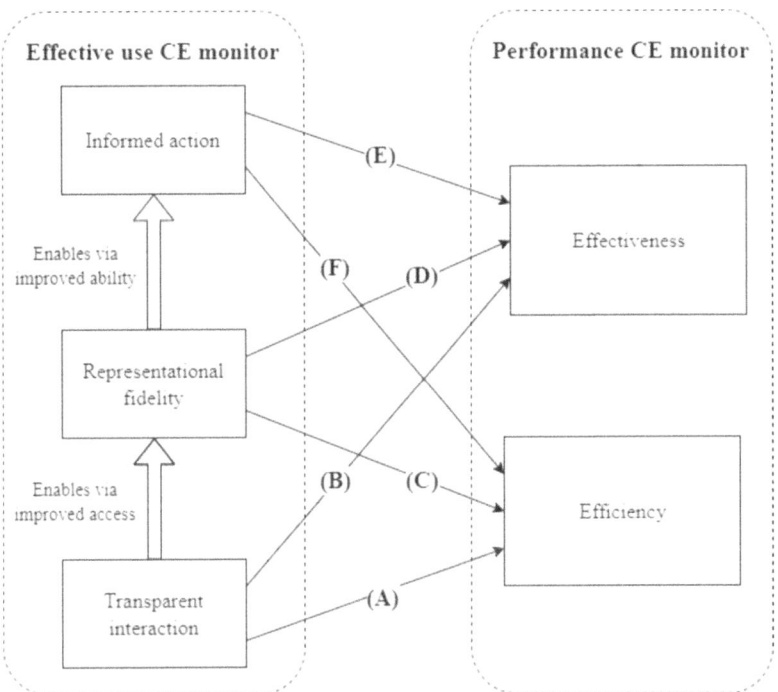

Fig. 1. Effective use CE monitor and its impact on performance. Framework adopted from Burton-Jones and Grange [9].

For transparent interaction, the openness of the monitor allows policymakers to efficiently demonstrate the added value of their policies to the public, thereby enhancing accountability and public trust in CE policies (A) [29]. Additionally, when policymakers engage with the monitor on a regular base, they are more likely to remain focused on CE information, indicators, and policy targets, leading to better integration of CE into policy departments and more accurate policy evaluation (B). For representational fidelity, our findings suggest that implementing a CE monitor should enhance policymakers' understanding of the CE domain by providing a centralized overview of CE information. This centralization leads to greater efficiency in consulting and researching information for policy development, as less time should be spent verifying sources and fidelity (C). Additionally, interviewees emphasized the importance of the monitor's ability to establish relationships between its indicators and provide insights into the system-thinking nature of CE. This ability of the monitor improves the effectiveness of addressing the complexity of CE in policymaking and enhances policymakers' understanding and knowledge capacity in the CE domain (D). For informed action, our interviews revealed that a CE monitor should enhance the effectiveness of integrating CE information into evidence-based policymaking. Conversely, poorly informed policy decisions can undermine the transition towards CE. We identified three potential use cases for informed action with the CE monitor. First, policymakers can use it to support their arguments and positions in policy meetings. Second, it can serve as evidence for evaluating existing policies. Third, it can provide input for the development of new policies (E). Although not explicitly mentioned in the interviews, an additional use case derived from Burton-Jones and Grange's [9] framework is that a CE monitor can facilitate informed action and increase the efficiency of policymaking by reducing the time spent correcting errors caused by poorly executed decisions (F). This mapping of desired effective use cases of the CE monitor clearly highlights the monitor's potential role in embedding CE into policymaking. However, as indicated by interviewee responses, the current lack of effective use of the monitoring systems impedes the monitor from fulfilling its intended purposes. Therefore, understanding the broader disturbances related to embedding CE in policymaking is crucial. This understanding can inform improvements to the monitor, enhancing its functionalities and overall effectiveness.

4.3 Identifying the Disturbances for Effective CE Policy Making

Völker et al. [3] highlight the importance of developing indicators to measure progress and inspire collective engagement with CE principles as a prerequisite for effective CE policy implementation. However, our interviews reveal that simply having a monitoring system is insufficient to embed CE into evidence-based policymaking. To better understand the challenges of integrating CE principles into policymaking, we reviewed existing literature to validate our problem situation (4.1) (Table 1).

The First Identified Issue (ISS1) Highlights the Long-Term Horizon of Circular Actions. Policymakers often operate within short-term electoral cycles and prioritize immediate results over long-term sustainability goals [30]. This is also true for CE policies, as their effects are not immediately visible, as CE is a transitional concept [7]. This mismatch between policy timeframes and the longer-term benefits of circularity initiatives makes it difficult for policymakers to commit to turning words into action and to gain the political support to allocate resources effectively [31].

The Second Identified Issue (ISS2) is the Fragmented Policy Mix and Uncoordinated Policy Efforts. The CE has since 2015 been high on the EU political agenda since the introduction of the Circular Economy action plan [10]. Numerous EU policy instruments and directives have been adopted to promote more sustainable resource use in areas such as waste management, product regulation, and chemicals policy [32]. However, as Kautto & Lazarevic [33] point out that these policies are developed and implemented within separate departments which sometimes have conflicting objectives and operate at various governance levels (EU, national, and regional). Without collaboration between departments and alignment of policies, there is a risk that these policies will be too narrowly focused and not effectively implemented, thus failing to achieve the transformative goals of CE [34].

The Third Identified Issue (ISS3) is the Absence of a Supportive Policy Agenda that Sets the Stage for Higher R-Strategies. The existing EU policy framework primarily concentrates on the end-of-life stage, aiming to reduce resource loss and enhance material circulation, through better recycling. This approach is marked by numerous waste-related directives and regulations that advocate for a "waste hierarchy" management system [2]. This focus primarily targets lower R-strategies (recycle, recover), while neglecting higher R-strategies (reduce, reuse and repair). This argument is reinforced by our interviews with policy makers. As one interviewee explained: *"Essentially, we provide advice based on* [waste management policy plans]. *These are the most substantive policy documents we receive, and everyone approaches them with great commitment. This topic really engages the consulting partners"*. Furthermore, Fitch-Roy et al. [35] found that current CE policies are merely incremental adjustments to existing policies within a linear resource system. These adjustments are insufficient to address the transformational nature and society-wide scale required to transition to a CE. As the priorities of the CE policy agenda remain focused on lower R strategies, embedding CE information into policy domains will have limited impact on improving material resource efficiency and effectivity.

The Fourth Identified Issue (ISS4) Highlights the Complex Nature of Implementing CE Principles. Implementing effective CE principles requires balancing environmental, economic, and social impacts while facilitating information flows across value chains, multiple lifecycles, and various societal actors [36]. The interactions among these actors create a complicated policy landscape that spans different parts of production and consumption cycles and risk direct or indirect rebound effects in different parts of the value chain. For instance, the introduction of reusable packaging directive might shift the impacts from the packaging producer to the reverse logistics and from the packaging production and waste treatment to the use phase. As one interviewee noted, *"You can look at CE from many different perspectives. I think that even among policymakers this understanding has not yet taken hold. Many policymakers are still not aware that you have to look at CE globally from a footprint perspective. You still have to convince or at least inform some policymakers."* Recent studies have shown that the absence of this systemic lifecycle perspective hampers the development of policies that stimulate CE in industries such as construction [37], food [38], and textiles [39]. Additionally, this complexity and lack of system focus has undermined effective policy development for material streams such as plastics [40] and nutrients [41].

The Fifth Identified Issue (ISS5) Underlines the Discrepancy Between Theoretical Frameworks and Practical Policy Tasks. Although this issue is not explicitly documented in the literature, our interviews revealed that policymakers experience a mismatch between the conceptual framework of the CE monitor developed by researchers and the practical boundaries of policy tasks and policy processes. As one interviewee explained: *"… But where improvement [of the monitor] might still be needed is better alignment with policy and policy goals. I think that initially, we just created a conceptual* *framework for the CE monitor based on consumption domains, and sometimes we tried to fit certain consumption domains into this framework just to meet the conceptual and theoretical requirements. This results in misalignment with policy"*. As a consequence, policy makers don't regularly access the monitor because it doesn't allow them to transparently interact with their policy needs. Therefore, a better alignment between research activities and policy tasks can enhance the embedding of CE in policymaking.

Table 1. Validated problem statement as the artifact of the Problem Analysis Echelon.

#	Problem element
ISS1	Circular actions have a long-term horizon beyond democratic terms
ISS2	Fragmented policy mix and uncoordinated policy efforts
ISS3	Absence of a supportive policy agenda that sets the stage for higher R-strategies
ISS4	Complex nature of implementing CE principles
ISS5	Discrepancy between theoretical frameworks and practical policy tasks

5 Outlook for Improvements for the Effective Use of the Monitor

To improve the effective use of a CE monitoring system and solving these identified issues of embedding CE in policy making, we draw four DRs and formulated four DP's that are informed and structured by our kernel theory TEU [9].

5.1 Design Requirements

For an IS to be effectively used, users must have direct access to the system's representations. As demonstrated in ISS4 and ISS5, users of the monitor struggle to extract meaningful information due to the complexity of CE, which necessitates a system thinking approach (ISS4), and the misalignment between theoretical frameworks and the information needs for policy tasks and processes (ISS5). Consequently, users are unable to interact transparently with the system. One approach to improve the transparent interaction is to adapt the system's surface structure. To address these challenges, we propose the first DR to enhance users' ability to interact transparently with the monitor:

DR1: The monitoring system should offer detailed, user-specific information on the CE to satisfy the specific information needs of intended users, thereby facilitating higher levels of transparent interaction. The second level in the hierarchy of TEU is representational fidelity, the extent to which a user can obtain a faithful representation of the underlying domain by the system. Because the tangible benefits of CE policies often materialize only in the distant future (ISS1), policymakers find it challenging to commit to these actions. This results in a lack of fidelity in the represented information of the monitor due to uncertainty about the future and a preference for immediate result. Therefore, to improve the user's fidelity in the representation of the CE domain in the monitor, users must have access not only to high-level macro indicators, which typically change only in the long term, but also to signal indicators [42]. These signal indicators provide more short-term feedback on the evolution of indicators and can serve as a proxy for assessing the immediate impact of policy decisions. This multi-layered approach improves the representational fidelity of the monitoring system by addressing the challenge of aligning the time dimensions of policy action and policy impact. We therefore propose the second DR:

DR2: The monitoring system should provide direct policy feedback to its users on the CE to improve the user's fidelity in the representation of the system. The final dimension of TEU is the informed action taken by users based on the information provided by the representations of the system. One common issue related to the lack of fidelity of the monitoring system and reducing policy makers ability to take informed actions is the complexity of understanding the concept of CE (ISS4) and aligning policy decisions over different departments (ISS2). The interviews revealed that policy makers often need guidance to unravel this complexity and translate the information from the monitor into policy decisions. Assisting policymakers in using and processing information from the CE can improve the fidelity of the monitoring system and enhance the coordination between policymakers' and their ability to formulate informed actions. We therefore propose the third DR:

DR3: The monitoring system should assist its users in translating information into knowledge, thereby improving their ability to make evidence-based decisions. Embedding CE principles in policymaking aims to shift the focus from end-of-life policies to those that improve resource efficiency and effective resource use in our society. However, as highlighted in ISS3, the lack of emphasis on higher R strategies in the current policy agenda means that policy makers don't have direct information needs to access the monitor, reducing the ability of the monitor to contribute to informed action for the formulation of higher R strategy policies. Therefore, we argue that in the absence of CE-stimulated policies, the CE monitoring system should not only provide indicators to evaluate existing policies but also focus on developing new indicators related to higher R-strategies. This can guide and shape the development of new policies, thereby enhancing informed action by policymakers and promoting the adoption of higher CE strategies. We therefore propose the fourth DR to improve the informed action of the CE monitor:

DR4: The monitoring system should prioritize providing information on higher R-strategies, shifting users' knowledge and focus from waste management and recycling to resource efficiency and effectiveness.

5.2 Design Principles

To illustrate the solvability of our problem statement in the context of the CE monitoring system, we formulate four Design Principles (DPs) (Fig. 2). We followed the recommendation of Chandra et al. [43] to formulate our DPs, who define DPs as prescriptive guidelines for developing or improving ISs.

DP1: Provide the monitoring system with a sequence of guidelines in order for users to navigate the system and find relevant information to satisfy their information needs. Assisting the user in finding the appropriate information in the system through guidelines enhances the user's transparent interaction with the system (DR1) and improve user's ability to access the surface structure of the monitor and translate the relevant information into actionable insights (DR3).

DP2: Organize the monitoring system in layers in order for users to easier find relevant information. By improving the system's deep structure with a range of indicators from macro to micro level, users can efficiently find the most relevant information for their short-term needs and increase their fidelity in the system (DR2), while maintaining an overall view of the CE priorities (DR4).

DP3: Establish relationships between different layers of information in order for users to better understand and process the represented information on the complex nature of CE in the system. Although monitoring systems summarize evidence on the state of a domain, it is difficult for users to translate the accessed information into knowledge and actionable decisions. By not only providing an overview of CE information, but also explaining the relationships between layers and indicators, DP3 addresses directly DR3 by enhancing users' ability to take informed actions and indirectly DR1 by improving users' transparent interaction with the system. **DP4: Prioritize information on the higher-level R-strategies in the monitoring system in order for users to shape policies that contribute to the overall objectives of the CE.** This DP directly address DR4. Therefore, the system's representation needs to focus on the key material properties of the CE, providing users with the affordances to make informed actions for the CE transition.

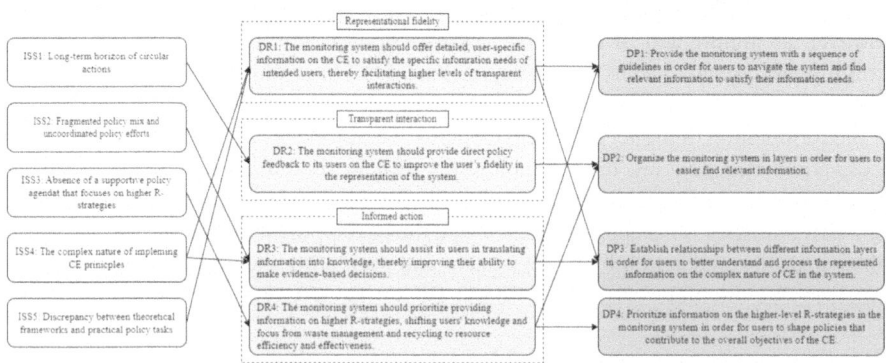

Fig. 2. Issues (ISS), design requirements (DR), and design principles (DP).

6 Discussion and Conclusion

This paper presents the second design cycle of our DSR project, dedicated to the design and development of a CE monitor for policymakers. Through 13 semi-structured interviews and a comprehensive literature review on the challenges of embedding CE in policymaking, we formulated and validated a problem statement comprising five key issues that hinder the integration of CE principles into evidence-based decision-making.

To address the ineffective use of the CE monitor, we derived four DRs and four DSs aimed at enhancing its effective utilization. This study contributes to the literature on policy monitoring frameworks and CE policymaking. Our interviews with the intended users of the monitor provided insights in the importance of understanding the problem space, specifically why current monitors are not effectively used and what their potential use cases are. We found that although policymakers value the existence of the CE monitor for potential applications—such as supporting arguments and positions in policy meetings, providing evidence for evaluating existing policies, and offering input for the development of new policies they currently underutilize the monitor. This underutilization is due to several issues in CE policymaking, including the long-term horizon beyond democratic terms, the fragmented policy mix, the absence of higher R-strategy policies, the complex nature of CE, and the discrepancy between theoretical frameworks and practical policy tasks. This research also makes two methodological contributions. By employing the eDSR method and developing the problem analysis echelon, we validated our problem statement, thereby providing design knowledge for further researchers. Future research can use these problem statements as a starting point to solve specific issues for CE policy making. Furthermore, our outlook for DRs and DSs offer preliminary guidelines for improving existing monitoring systems, which can be validated in future research. These insights emphasize the need for user-friendly interfaces, integration with existing policy frameworks, emphasis on higher R-strategies and short term policy feedback tailored to the specific information needs of policymakers. Future research should validate these DPs and implement them in existing artifacts to ensure that the design of the CE monitor aligns better with the identified requirements, thereby bridging the gap between theoretical CE frameworks and their practical application in policymaking.

Acknowledgments. M. Pauwels is grateful for KU Leuven's funding provided by VITO (Grant No: 3E230665). R. H. Reich is grateful for the internal C2-funding of KU Leuven (Grant No: 3E210013). L. Alaerts is grateful for the financial support received from the Flemish administration via CE Center (Steunpunt Circulaire Economie) This publication contains the opinions of the authors, not those of the Flemish administration. The Flemish administration will not carry any liability with respect to the use that can be made of the produced data or conclusions.

Disclosure of Interests. The authors have no competing interests to declare that are relevant to the content of this article.

References

1. Reichel, A., De Schoenmakere, M., Gillabel, J., Martin, J., Hoogeveen, Y.: Circular economy in Europe: Developing the knowledge base. Eur Environ. Agency Rep. **2**, 2016 (2016)
2. Milios, L.: Advancing to a circular economy: three essential ingredients for a comprehensive policy mix. Sustain. Sci. **13**, 861–878 (2018). https://doi.org/10.1007/s11625-017-0502-9
3. Völker, T., Kovacic, Z., Strand, R.: Indicator development as a site of collective imagination? The case of European commission policies on the circular economy. Cult. Organ. **26**, 103–120 (2020). https://doi.org/10.1080/14759551.2019.1699092
4. Recker, J.: Toward a design theory for green information systems. In: 2016 49th Hawaii International Conference on System Sciences (HICSS), pp. 4474–4483. IEEE (2016)

5. Oliver, K., Innvar, S., Lorenc, T., Woodman, J., Thomas, J.: A systematic review of barriers to and facilitators of the use of evidence by policymakers. BMC Health Serv. Res. **14**, 2 (2014). https://doi.org/10.1186/1472-6963-14-2
6. Alaerts, L., Reich, R.H., Pauwels, M., Van Acker, K.: Embedding a circular economy monitor in public administration (2024)
7. Alaerts, L., et al.: Towards a more direct policy feedback in circular economy monitoring via a societal needs perspective. Resour. Conserv. Recycl. **149**, 363–371 (2019). https://doi.org/10.1016/j.resconrec.2019.06.004
8. Tuunanen, T., Winter, R., vom Brocke, J.: Dealing with complexity in design science research: a methodology using design echelons. Manag. Inf. Syst. Q. **48**, 427–458 (2024)
9. Burton-Jones, A., Grange, C.: From use to effective use: a representation theory perspective. Inf. Syst. Res. **24**, 632–658 (2013). https://doi.org/10.1287/isre.1120.0444
10. European Commission: Closing the loop—An EU action plan for the circular economy (2015)
11. Alberich, J.P., Pansera, M., Hartley, S.: Understanding the EU's circular economy policies through futures of circularity. J. Clean. Prod. **385**, 135723 (2023)
12. Hartley, K., Schülzchen, S., Bakker, C.A., Kirchherr, J.: A policy framework for the circular economy: Lessons from the EU. J. Clean. Prod. **412**, 137176 (2023)
13. Esty, D., Rushing, R.: The promise of data-driven policymaking. Issues Sci. Technol. **23**, 67–72 (2007)
14. Mügge, D.: Studying macroeconomic indicators as powerful ideas. J. Eur. Publ. Policy **23**, 410–427 (2016). https://doi.org/10.1080/13501763.2015.1115537
15. Van Veenstra, A.F., Kotterink, B.: Data-driven policy making: the policy lab approach. In: Parycek, P., et al. (eds.) EPart 2017. LNCS, vol. 10429, pp. 100–111. Springer, Cham (2017). https://doi.org/10.1007/978-3-319-64322-9_9
16. Diran, D., Van Veenstra, A.F.: Towards data-driven policymaking for the urban heat transition in The Netherlands: barriers to the collection and use of data. In: Viale Pereira, G., et al. (eds.) EGOV 2020. LNCS, vol. 12219, pp. 361–373. Springer, Cham (2020). https://doi.org/10.1007/978-3-030-57599-1_27
17. Diran, D., Hoekstra, M., Van Veenstra, A.F.: Applications of data-driven policymaking in the local energy transition: a multiple-case study in the Netherlands. In: Krimmer, R., Rohde Johannessen, M., Lampoltshammer, T., Lindgren, I., Parycek, P., Schwabe, G., Ubacht, J. (eds.) EPart 2022. LNCS, vol. 13392, pp. 55–72. Springer, Cham (2022). https://doi.org/10.1007/978-3-031-23213-8_4
18. Head, B.W.: Three lenses of evidence-based policy. Aust. J. Public Adm. **67**, 1–11 (2008). https://doi.org/10.1111/j.1467-8500.2007.00564.x
19. Janssen, M., Helbig, N.: Innovating and changing the policy-cycle: policy-makers be prepared! Gov. Inf. Q. **35**, S99–S105 (2018)
20. Head, B.W.: Reconsidering evidence-based policy: key issues and challenges. Policy Soc. **29**, 77–94 (2010). https://doi.org/10.1016/j.polsoc.2010.03.001
21. Young, I., et al.: Experiences and attitudes towards evidence-informed policy-making among research and policy stakeholders in the C Anadian agri-food public health sector. Zoonoses Public Health **61**, 581–589 (2014). https://doi.org/10.1111/zph.12108
22. Bangert, H.: Japan's circularity. A panorama of Japanese policy innovation technology and industry contributions towards achieving the Paris agreement. Report, EU-Japan Centre for Industrial Cooperation (2020)
23. Geng, Y., Doberstein, B.: Developing the circular economy in China: challenges and opportunities for achieving "leapfrog development". Int J Sust Dev World **15**, 231–239 (2008). https://doi.org/10.3843/SusDev.15.3:6
24. Magnier, C., et al.: Ten key indicators for monitoring the circular economy. Environmental Information Department, Ministry of the Environment, Energy and Marine

Affairs, France (2017). http://www.statistiques.developpement-durable.gouv.fr/publications/p/2669/1539/10-indicateurs-cles-suivi-leconomie-circulaire-edition-2017.html. Accessed 2 Oct 2018
25. Potting, J., Hanemaaijer, A., Delahaye, R., Ganzevles, J., Hoekstra, R., Lijzen, J.: Circular economy: what we want to know and can measure. Framework and baseline assessment for monitoring the progress of the circular economy in the Netherlands, p. 92 (2018)
26. Mayer, A., Haas, W., Wiedenhofer, D., Krausmann, F., Nuss, P., Blengini, G.A.: Measuring progress towards a circular economy: a monitoring framework for economy-wide material loop closing in the EU28. J. Ind. Ecol. **23**, 62–76 (2019). https://doi.org/10.1111/jiec.12809
27. Vom Brocke, J., Winter, R., Hevner, A., Maedche, A.: Special issue editorial–accumulation and evolution of design knowledge in design science research: a journey through time and space. J. Assoc. Inf. Syst. **21**, 9 (2020)
28. Weber, R.: Ontological foundations of information systems. (No Title) (1997)
29. Matheus, R., Janssen, M., Maheshwari, D.: Data science empowering the public: data-driven dashboards for transparent and accountable decision-making in smart cities. Gov. Inf. Q. **37**, 101284 (2020)
30. Voß, J.-P., Smith, A., Grin, J.: Designing long-term policy: rethinking transition management. Policy. Sci. **42**, 275–302 (2009)
31. Friant, M.C., Vermeulen, W.J., Salomone, R.: Analysing European union circular economy policies: words versus actions. Sustain. Prod. Consum. **27**, 337–353 (2021)
32. Hartley, K., van Santen, R., Kirchherr, J.: Policies for transitioning towards a circular economy: expectations from the European Union (EU). Resour. Conserv. Recycl. **155**, 104634 (2020)
33. Kautto, P., Lazarevic, D.: Between a policy mix and a policy mess: policy instruments and instrumentation for the circular economy. In: Handbook of the Circular Economy, pp. 207–223. Edward Elgar Publishing (2020)
34. Weber, K.M., Rohracher, H.: Legitimizing research, technology and innovation policies for transformative change: combining insights from innovation systems and multi-level perspective in a comprehensive 'failures' framework. Res. Policy **41**, 1037–1047 (2012)
35. Fitch-Roy, O., Benson, D., Monciardini, D.: Going around in circles? Conceptual recycling, patching and policy layering in the EU circular economy package. Environ. Polit. **29**, 983–1003 (2020). https://doi.org/10.1080/09644016.2019.1673996
36. Zeiss, R., Ixmeier, A., Recker, J., Kranz, J.: Mobilising information systems scholarship for a circular economy: review, synthesis, and directions for future research. Inf. Syst. J. **31**, 148–183 (2021). https://doi.org/10.1111/isj.12305
37. Yu, Y., Junjan, V., Yazan, D.M., Iacob, M.-E.: A systematic literature review on Circular Economy implementation in the construction industry: a policy-making perspective. Resour. Conserv. Recycl. **183**, 106359 (2022)
38. Liu, Y., Wood, L.C., Venkatesh, V.G., Zhang, A., Farooque, M.: Barriers to sustainable food consumption and production in China: a fuzzy DEMATEL analysis from a circular economy perspective. Sustain. Prod. Consum. **28**, 1114–1129 (2021)
39. Peleg Mizrachi, M., Tal, A.: Regulation for promoting sustainable, fair and circular fashion. Sustainability. **14**, 502 (2022)
40. Soenmez, C., Venkatachalam, V., Spierling, S., Endres, H.-J., Barner, L.: Environmental potential of recycling of plastic wastes in Australia based on life cycle assessment. J. Mater. Cycles Waste Manag. **26**, 755–775 (2024). https://doi.org/10.1007/s10163-024-01901-1
41. Yuille, A., et al.: UK government policy and the transition to a circular nutrient economy. Sustainability. **14**, 3310 (2022)

42. Moraga, G., et al.: Circular economy indicators: what do they measure? Resour. Conserv. Recycl. **146**, 452–461 (2019). https://doi.org/10.1016/j.resconrec.2019.03.045
43. Chandra, L., Seidel, S., Gregor, S.: Prescriptive knowledge in IS research: conceptualizing design principles in terms of materiality, action, and boundary conditions. In: 2015 48th Hawaii International Conference on System Sciences, pp. 4039–4048. IEEE (2015)

Exploring Data Altruism as Data Donation: A Review of Concepts, Actors and Objectives

Dwayne Ansah(✉) and Iryna Susha

Copernicus Institute of Sustainable Development, Utrecht University, Princetonlaan 8a, 3584 CB Utrecht, The Netherlands
{d.o.ansah,i.susha}@uu.nl

Abstract. This paper reviews data altruism in the emerging academic literature by collecting and analyzing conceptually similar terms, identifying key actors, and delineating the objectives of general interest underlying this novel form of voluntary data sharing. Drawing from a wide array of disciplines including computer science, social science, law, and medicine and bioethics, we discover frequent comparisons with the notions of data donation, data crowdsourcing, data philanthropy, and data solidarity. In our observations of current definitions and understandings of data altruism and these related terms, we draw attention to analyzing their salient similarities and differences. Our analysis of actors and roles illustrates a wide variety of players envisioned to participate in data altruism. Our examination of different types of data discussed in the literature suggests the data altruism value chain proposition is largely promissory, with limited empirical evidence supporting specified objectives of general interest. This review contributes to a refined understanding of data altruism as a formal 'data donation' institutionalization process in the European digital space and highlights its implications for future research.

Keywords: data altruism · data donation · EU strategy for data · data sharing · data for good

1 Introduction

In the digital society, data is considered a resource for innovation and decision support. While the private sector enthusiastically explores the potential of data analytics within the burgeoning data-driven economy [1], the transformative power of data extends far beyond commercial applications. There is a growing belief that data can be harnessed for the betterment of society. Nowadays, numerous actors are actively engaged in political activities to use data to produce public goods and services, exemplifying the "data for good" movement [2–5].

Among these actors, the EU stands out for its adoption of the Data Governance Act (DGA) on 16 May 2022, thus creating new regulatory possibilities in the realm of EU data governance. This act introduces, among other things, data altruism, an emerging framework for regulating and facilitating the non-commercial, voluntary sharing of data for objectives of general interest, such as combating climate change, improving traffic and public health (Recital 45 DGA).

The introduction of data altruism as a policy instrument has drawn fresh scholarly attention. A cursory glance at these initial academic articles leads to the discovery of contest as to the "novelty" of data altruism. These scholars argue that data altruism is not entirely unprecedented, as it has parallels in the longstanding discourse surrounding data donation [6, 7], referencing semantically similar concepts such as data philanthropy and data solidarity. Rather than engaging in the debate solely on its novelty, we contend that the introduction of data altruism presents an opportunity for reflection on its distinctive characteristics, thereby laying the groundwork for future research endeavors.

To understand or evaluate the distinctiveness of data altruism, it is necessary to embed the concept within relevant academic discourses. The literature has indeed connected it closely with the notion of data donation, suggesting a comparative inquiry as a logical starting point for exploration. However, a closer examination reveals a significant gap: the absence of any mention of "donation" within the legislative text of the DGA. This absence emphasizes the necessity of probing not just what is the relationship between data altruism and data donation, but also what is useful or desirable in drawing this comparison. Hence, our research question: *How can data altruism be understood as data donation?* In this light, we first analyze data altruism as a new regulatory framework that is part of ongoing EU data governance reform efforts, before drawing grounded and constructive comparisons to the data donation discourse.

Then, in order to not only rely on a comparative definition of data altruism, we adopt a pragmatic approach to understanding the phenomenon, focusing on how we can make sense of data altruism in terms of *who is involved* and *what types of data are shared for which objectives of general interest.*

Understanding the distinctiveness of data altruism holds both societal and scientific significance. Societally, conceptual clarity on the novelty of data altruism will aid in the data literacy of the actors encouraged to share their data in the general interest. From a policy perspective, clear conceptualization supports policymakers and regulators as they navigate this emerging field of data governance. Scientifically, the notion of novelty fosters innovation and theoretical advancement in the literature. Namely, our review addresses the fragmentation of the research field populated by several related concepts and thereby fosters a more contextualized understanding of data altruism. Furthermore, this paper provides the first review of the data altruism literature.

The rest of the paper is structured as a series of delineations, discussions, and deliberations. Section 2 positions data altruism as a regulatory framework as part of ongoing EU data governance reform efforts. Subsequently, we discuss the methodological approach of the inquiry. Section 4 details the results of the analysis. In the discussion, we summarize and integrate our findings. Finally, in the conclusion and suggestions for future research, we draw together a plausible conceptualization of the distinctiveness of data altruism and explain its implications for the research agenda emerging in the nascent field.

2 Data Altruism as a Regulatory Framework

Data altruism, as defined in Article 2 (16) DGA, "means the voluntary sharing of data on the basis of the consent of data subjects to process personal data pertaining to them or permissions of data holders to allow the use of their non-personal data without seeking or receiving a reward that goes beyond compensation related to the costs that they incur when they make their data available for objectives of general interest as provided for in national law, where applicable, such as healthcare, combating climate change, improving mobility, facilitating the development, production and dissemination of official statistics, improving the provision of public services, public policy making or scientific research purposes in the general interest." Personal data refers to information concerning an identified or identifiable natural person, such as name, online identifiers or other information related to their identity (Art. 2 (3) DGA and Art. 4 (1) GDPR). Non-personal data refers to all data other than personal data.

Furthermore, the DGA introduces a public national register of recognized data altruism organizations (RDAOs). RDAOs are a new type of data intermediary expected to lead to the creation of sufficiently-sized data pools, which are intended to fuel data analytics and machine learning for objectives of general interest (Recital 45). Each Member State designates their own competent authority for the registration of RDAOs (Art. 23 DGA). RDAOs may be established and, thus, registered in any Member State (Art. 19 DGA), which promotes harmonization and ease of cross-border data sharing within the EU (Recital 46). RDAOs must comply with general requirements for registration (Art. 18 DGA), transparency requirements (Art. 20 DGA), a to-be-established rulebook (Art. 22 (1) DGA), and specific safeguards to protect the rights and interests of data subjects and holders (Art. 21 DGA).

Recital 46 provides clarification that while RDAOs have the duty to carry out data altruism activities (Art. 18 (a) DGA), the process of formal registration as a recognized entity under the DGA should not be a mandatory requirement for conducting such activities. This regulatory nuance means that organizations can still carry out data altruism activities without needing to be officially registered as an RDAO. Following Art. 2 (16) DGA, all data altruism activities relate to facilitating voluntary data sharing between (1) potential data users and (2) the potential individuals or entities providing data (from here on: data donors or donors), for objectives of general interest.

The potential data user of voluntarily shared data must be a natural or legal person with lawful data access and the right to non-commercially use that data for its stated objective(s) of general interest. It is possible that the RDAO doubles as data user if it carries out "data altruism activities through a structure that is functionally separate from its other activities" (Art. 18 (d) DGA). In other cases, RDAO functions as an intermediary allowing third parties to utilize data (Art. 21 DGA).

To assist potential data donors to easily identify and trust RDAOs, a common recognizable logo is introduced (Recital 47). The RDAO is tasked with providing tools for obtaining and easily withdrawing consent, in case of personal data, or permission, in case of non-personal data, from data donors (Art. 21 DGA). Article 25 DGA introduces a new, to-be-established European data altruism consent form to allow the collection of consent or permission in a uniform format, regardless of Member State.

In the broader context of EU policies concerning its data economy, data altruism is just one of many instruments introduced to steer it. The European Strategy for Data, published in February 2020, comprises a suite of legislative instruments: DGA, Data Act, Implementing act on high value datasets, Digital Markets Act, and Digital Services Act. Each legislation focuses on different aspects of the EU data economy.

3 Methodology

Drawing methodological inspiration from a pragmatic framework for guiding and evaluating literature reviews, we prioritized rigor, relevance and methodological coherence [8] in our exploration of data altruism. This involved conducting a developmental review to identify (under)explored research areas [9–11].

Concretely put, our data collection followed a two-phase approach. In Phase 1, we conducted an in-depth data extraction procedure from ten academic articles on data altruism after screening for relevance. Given the field's emerging nature, our search yielded a modest fourteen academic articles using the search term "data altruism." Four articles were excluded: two articles were inaccessible, one was written in the Russian language, and another had in its abstract, coincidentally, one sentence that ended with the "data." followed by a sentence that started with "Altruism". The ten remaining articles were used for the data extraction procedure. We used NVivo 14 to iteratively develop a coding scheme [11] with the intent to systematically extract and analyze (1) key terms to guide our attempt to map related concepts, (2) the actors involved in data altruism, following our ad hoc distinction between primary actors (i.e., those engaged as either data donor, RDAO or data user) and institutional actors (i.e., all other actors defined in the regulatory framework, see Sect. 4.2), and (3) the types of data voluntarily shared and their objectives of general interest based on the non-exhaustive list of examples in the DGA. In Phase 2, we conducted a broad scan of the literature covering the related concepts discovered in Phase 1: data donation, data philanthropy, data solidarity, and data crowdsourcing.

We searched for literature using Scopus and Web of Science using the terms to search within title, abstract, and article keywords. This resulted in a sample of 843 articles. Google Scholar was deliberately not used as a search database due to concerns about its limited coverage and lack of functionalities to ensure reproducibility of search results [12]. 277 duplicate articles were removed. The first author screened the articles to identify written definitions of and concepts surrounding voluntary data sharing for objectives of general interest, excluding those written in languages other than English, and grey literature to maintain focus on mapping the state of the art within academic discourse. The final literature sample includes 167 relevant papers written in English from 2010 until August 2023.

We conducted an analysis for mapping current definitions and understandings surrounding data altruism, and content analysis aimed at clarifying who is involved in data altruism and what type of data is voluntarily shared for which objectives of general interest.

4 Results

Data altruism is a new concept in the academic literature, with all articles in our sample having been published after 2020. The concept has been discussed in a wide variety of fields, as evident from the diversity of journals found in our sample: computing [13–16], social science [1, 17], law [1, 18], and medicine and bioethics [6, 7].

4.1 What are Current Definitions and Understandings of Data Altruism

Ongoing debates surrounding terminology reinforce the need for clarity and coherence regarding the distinctiveness of data altruism [6, 7]. To address this, we examine the relationship between data altruism [6, 7, 16, 19] and its related concepts. This exploration is grounded in an analysis of the literature collected in Phase 1 of our study and includes related concepts such as data donation [20–23], data philanthropy [20, 23–30], data crowdsourcing [31–35] and data solidarity [36–42].

The data altruism literature often interprets the data altruism regime as an opportunity to regulate data donation [16, 17, 19], sometimes leading to an equivocation between data altruism activities (Art. 18 (a) DGA) and data donation. We remind the reader that the DGA omits the term data donation. Unlike cases where personal data is used as counter-performance or substitute for payment [43], data altruism activities concern actors willfully sharing without the motivation of a direct benefit from their efforts [44]. Data altruism is distinct from capitalist notions like surveillance capitalism [45], offering a framework where data sharing is rooted in altruistic principles, rather than purely economic motives. The goal of the data altruism provisions is to create trusted tools for voluntary data sharing and to assure individuals and organizations donating data that it will be handled in accordance with EU values and principles [46]. Thus, data altruism concerns the non-commercial part of the EU data economy.

We propose to understand data altruism as a unique form of voluntary data sharing, agnostically related to data donation and other related terms. Data altruism activities involve the generation and collection of new data or the sharing of existing data with the express aim of serving objectives of general interest [17, 19]. From this distinct perspective, data altruism has been considered a legal [17], governmental [17], informational [17], methodological [19], social [19] and data-driven [19] innovation, requiring regulation to validate and standardize appropriate voluntary data sharing practices to ensure that data is used exclusively in the general interest [19].

Data donation is conceptually similar to data altruism, as in that it involves the practice of voluntarily sharing data for a societal cause, without expecting immediate or tangible benefits in return [20–23]. Thus, distinctive features of data donation are the voluntary motive of the donor and the prosocial purpose of the intended use of the donated resource. Appropriately, the data altruism literature lays an explicit connection to data donation [6, 7, 16, 19]. However, data donation understood as a legal instrument suggests that the data subject or holder transfers ownership of their data to another party, which is explicitly not the case in the European Strategy for Data where data protection and privacy are viewed as fundamental rights and, thus, cannot be contracted or transferred away as property [6, 7]. Instead, the notions of consent and permission suggest a more relational approach to data altruism. In sum, data donation operates independently of data altruism's legal and ownership implications.

Data philanthropy refers to private firms voluntarily sharing (proprietary) data in the public interest [20, 23–30], effectively serving as a corporate manifestation of data donation. Notably, some definitions of data philanthropy explicitly include individuals alongside private companies as donor [24, 25]. The data altruism literature frequently compares and contrasts data philanthropy [1, 6, 7]. Although the users of philanthropic data are largely unspecified, they may include public sector actors, international organizations, and industry actors [25]. Importantly, some scholars define data philanthropy as not limited to data donation and include all other related assets required to utilize donated data, like technology and expertise [26]. This latter view is particularly relevant for understanding data altruism activities as it emphasizes the role of the data user.

Data crowdsourcing is the process of collecting, generating or obtaining data by soliciting contributions from a large and diverse group of individuals, typically via the internet [31]. This large-scale collaboration overlaps with the concept of donation-based crowdsourcing for charitable causes [32, 33]. Participants in online donation-based crowdsourcing projects often also do so with altruistic motivations [34, 35]. This concept is also referred to in the data altruism literature [13, 17, 19], supporting its relevance as a voluntary data sharing practice other than data donation or as counter-performance.

Data crowdsourcing overlaps with data altruism in two important ways. First, it has been described as a new way for patients to participate in the advancement of personalized medicine, aligning with the EU's objectives of general interest [19]. Secondly, similar to how the EU regulates and encourages voluntary data sharing through data altruism provisions, the US has a regulatory regime for bottom-up data generation by non-governmental entities for scientific research purposes – another stated EU objective of general interest – under the US Crowdsourcing and Citizen Science Act (15 USC 3724) [17].

Notably, scholarly discussions on data crowdsourcing occur outside the data donation discourse. It is a key concept in the big data ecosystem literature [47, 48]. Herein, data crowdsourcing is understood as the collection of massive datasets typically associated with big data, which involves datasets having large and complex structures [31, 49]. Within this literature, data crowdsourcing is sometimes framed in terms of its implications for the EU data economy [47, 48]. While the big data ecosystem literature recognizes the impact of big data beyond commercial applications [48], it falls short of providing a detailed exploration of the "for good" aspect.

Despite its significance, data crowdsourcing's terminology is absent from the DGA. Likewise, the big data ecosystem's terminology is also not found in the DGA. For instance, in its literature we find terms like "data suppliers" are used to describe the roles of data subjects and data holders [47]. This mirrors the lexical confusion seen with the term "data donation," which prompted our investigation.

Data solidarity refers to the practice of making data processes visible and accessible for the benefit of the public good [36]. It is a data governance principle that emphasizes shared traits and values among parties involved in data sharing for good, particularly relevant in contexts like research biobanks and healthcare [37–39, 42]. Furthermore, data solidarity explicitly challenges hierarchical data structures and neoliberal market forces by emphasizing democratic norms that empower communities to produce collective benefits [36]. Proponents argue that possible harm caused by withholding health data could outweigh the risks associated with its use [41, 42] and propose that research with minimal risks should not necessarily require explicit consent [40]. Data solidarity aims to promote a more equitable distribution of both the risks and benefits associated with data sharing, including data donation. While data solidarity has its merits, it also carries risks, such as compromising individuals' self-determination and potentially contributing to profiling [42]. Thus, the data solidarity literature acknowledges the inherent trade-offs involved in prioritizing either data sharing for objectives of general interest or individual data sovereignty also found in the data donation literature [22], a notable omission in the data altruism literature.

Furthermore, the data solidarity literature provides constructive insights by introducing relevant principles that should underlie the various stages of data donation (e.g., data processing, data sharing, data use) and are, arguably, part and parcel of the institutional actors' data governance contribution to the regime. These principles are derived from the concept of data justice, which involves fair evaluation of data resources and structures to prevent injustices, considering instrumental, procedural, and distributive aspects [38, 50]. This is exemplified in, for example, the transparency requirements and the introduction of the European data altruism consent form.

Based on our discussion of related concepts thus far, understanding data altruism presents a challenging task. We discover that the relationship between data altruism and related terms is contingent, fuzzy, creative, intuitive, and ambiguous [51]. Perhaps unsurprisingly, the meaning of data altruism will vary depending on the context. For example, someone working within a scientific research paradigm might view data altruism as the selfless sharing of data for the greater good of advancing knowledge, while a corporation might, following the data philanthropy line of thinking, consider it a strategic opportunity to enhance corporate social responsibility or improve public relations. Those emphasizing data altruism as a voluntary, prosocial act, often equate it to data donation, which becomes shorthand for sharing data voluntarily for public interest purposes without expecting a reward. Thus, data donation facilitates an intuitive, if procrustean, understanding of data altruism.

While the literature may conflate these terms, we stress the importance of recognizing data altruism as a formal regulatory process geared towards the general interest as defined in the DGA. Whereas data altruism shares common elements with the other concepts, it also represents a distinct and developing phenomenon with its own unique characteristics and objectives. Trust is a central feature of the data altruism regime, particularly evident in the establishment of the new trusted, not-for-profit intermediary RDAO, which adheres to EU values and standards. Furthermore, trust is facilitated by the DGA's introduction of mechanisms such as the common logo and transparency requirements. Importantly, participation in this regime is entirely voluntary; not-for-profit organizations carrying out data altruism activities have the opportunity to opt in through an official, public registration process.

4.2 Which Actors are Involved in Data Altruism and in What Roles?

As a regulatory framework, the data altruism regime involves not only the primary actors but includes a network of institutional actors shown in Fig. 1.

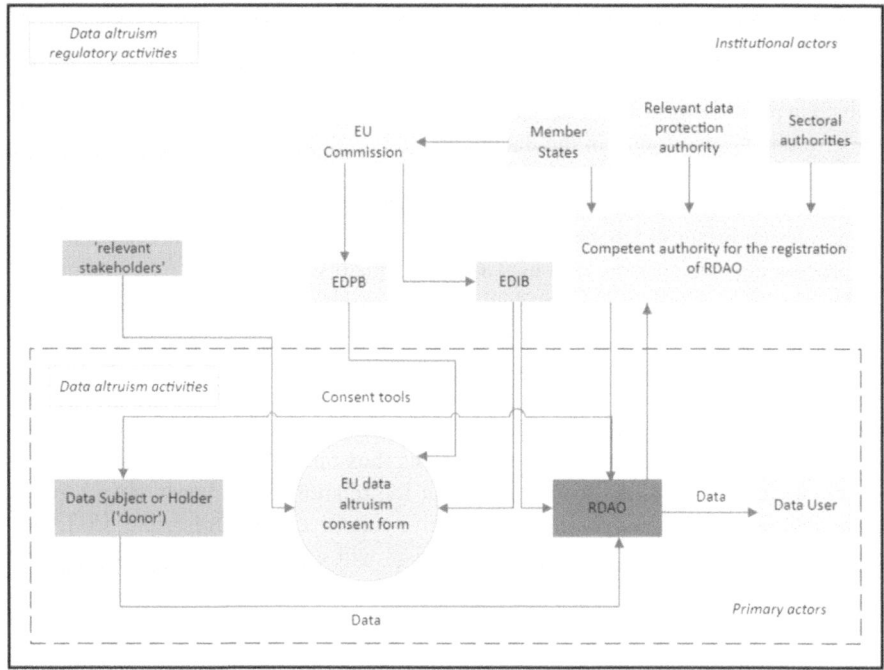

Fig. 1. Our schematic representation of actors in data altruism (regulatory) activities

Primary actors perform at either an individual or collective level. The literature mentions individual actors engaging in the donor role as citizen [6, 7, 17, 19], patient [6, 7, 19], researcher [16, 17, 19], citizen scientist [17], customer [19] and as deceased person [18]. Notably, individuals are the only actors who may give consent to sharing their personal data, embracing the Self-Sovereign Identity (SSI) paradigm [13, 14]. Individuals are more willing to share data with research organizations and governmental organizations than to firms [29], although this willingness is mediated by highly divergent national sensitivities [6]. Individuals cannot participate in the data altruism regime as intermediary (Art. 18 DGA) and there is also no mention of an individual functioning as data user, although this is legally allowed (Art. 2 (9) DGA).

Collective actors may perform the roles of donor, RDAO, and user. As donors, we find collective actors such as businesses [1, 15–17, 19] and citizen science projects [19]. RDAOs, regarded as data controllers (Art. 4 (7) GDPR), typically determine specific objectives of general interest and methodologies [18]. The literature provides ample plausible ante litteram examples of collective organizations across various domains carrying out data altruism activities: authorization servers [13], civil-society-led data collaboratives [19], research organizations [18], (government-led) citizen science projects

[19], the French Health Data Hub [6], the Danish research-focused system [6] and the German insurance data for health research scheme [6]. Voluntary registration as RDAOs offers potential advantages to these collective organizations, such as alignment with EU regulatory guidance and the opportunity to display the common logo, fostering trust, and encouraging voluntary data sharing. However, these actors may be deterred from registering as an RDAO due to the additional bureaucratic burdens [6, 18] and to-be-determined fines in case of noncompliance with the DGA [18]. Furthermore, the introduction of RDAO adds yet another layer of complexity to the existing data protection legal framework, which may lead to adverse effects, such as data losses [6], time delays [6] and increased legal uncertainty as to the permitted use of data in innovation projects [18].

Additionally, collective data users are categorized into four groups: research organizations [17–19], NGOs [18, 19], government organizations [17] and media organizations [17]. Under NGOs we identified, among others, advocacy groups. Under government organizations we identified national players, such as the public prosecutors' office, as well as EU-level players, such as the European Committee of the Regions. Although not explicitly discussed in the literature, commercial entities are also legally allowed to function as data users. In this sense, some researchers warn that there is a plausible business model risk, namely that companies will sell datasets obtained through the data altruism mechanism to other entities [6], thus bypassing the intended objectives of general interest.

Institutional actors are exclusively collective and engage in data altruism as either state or non-state actors. According to the DGA, at the state level, entities like the European Commission, the European Data Protection Board (EDPB), and the newly established European Data Innovation Board (EDIB) are involved. The EDIB is responsible for crafting guidelines to steer the European Strategy for Data forward [19]. The oversight of data altruism activities by the EDIB underscores its significance within the European Strategy for Data framework. Nationally, existing systems for 'data donation' vary significantly (e.g., Denmark, France and Germany [6]) and the literature does not discuss if and how these existing systems will relate to data altruism. Furthermore, relevant data protection authorities are tasked with preventing misinformation [17]. We identify plausible sectoral authorities to include environmental protection authorities [17] and proposed health data access bodies in the current European Health Data Space proposal, which will oversee data altruism activities related to health data and collaborate with competent authorities to supervise RDAOs [6].

Non-state actors, referred to as 'relevant stakeholders' in the DGA, include oversight actors and technology providers. Oversight actors are envisioned to oversee the quality of voluntarily shared data, promote digital literacy, and democratize data access for marginalized groups [6]. The DGA includes ethics councils or boards as part of the data altruism oversight mechanisms (Recital 46). Such actors may plausibly include patient groups [6], practitioners [19], researchers [19], and civil society associations [6, 19]. Conversely, technology providers play a more prominent role in the literature than highlighted in our discussion of the DGA in Sect. 2. Despite receiving limited attention, providers of personal data services (PDS) like the open-sourced SOLID project, facilitate the user-side distribution of personal data, thereby empowering individuals to exercise control over and enabling granular-level data sharing across various data silos [13–15]. Regarding research organizations, data-altruism-compliant software developers are mentioned in the literature [18]. Lastly, the Semantic Web Community is envisioned to provide tools to support competent data altruism authorities in their responsibility to keep accurate records of RDAOs' processing activities and help format the new European data altruism consent form into a machine-readable consent form [15, 18].

4.3 What Type of Data may be Voluntarily Shared and for Which Objectives of General Interest?

Like the DGA, the data altruism literature uses the non-personal/personal distinction. Non-personal data has been especially under-researched in the literature and merely includes mentions of environmental data [13, 17] and synthetic biometric data [16]. On the other hand, personal data has received significant attention. We broadly find two types of personal data: health data [6, 7, 14, 16, 17, 19] and user data [13, 14, 16, 19]. Health data includes electronic health records [6, 7, 14, 17, 19], physical activity data [14], biometric data [13] and genetic data [7]. User data can be divided into two categories, namely user-devices generated data (e.g., photographs, smart home data, geolocation data) and user-interaction generated data (e.g., financial transactions, telecom data) [18].

The line between distinct types of personal data is, at times, blurry. For example, where physical activity data generated by self-tracking apps would prima facie be considered user data, it also fits into the category of health data since it reveals information about an individual's health. Furthermore, transformations are possible from personal to non-personal data. This includes the production of synthetic data, where non-personal data is artificially generated from sufficiently anonymized data [7, 18] as to no longer relate to a specific individual. Reidentification presents a foreseeable risk. Another such transformation occurs after an individual's death: all their personal data becomes non-personal, allowing for a niche form of post-mortem data donation [18].

The literature largely shows the voluntary sharing of various data types for non-descript objectives of general interest, mentioning broad labels such as 'community interest' [14, 19], 'social good' [13] and 'common good' [15]. Only a few specific purposes are identified in the literature: environmental data is theoretically expected to help in improving mobility [13] and has demonstrably informed public policy making [17]; health data is mentioned as having contributed to healthcare advancements [19]; and, most popularly, a wide variety of personal and non-personal data have been shared for scientific research purposes in the general interest [6, 13, 16, 18, 19].

5 Discussion

Our review contributes to the existing body of literature by providing a comprehensive examination of current definitions and understandings of data altruism in the emerging scholarly literature in computer science, social science, law, and medicine and bioethics. Furthermore, we identified and delineated relevant actors, which types of data are expected to be voluntarily shared between them, and which objectives of general interest are mentioned in the literature thus far. Our review is explicitly not a legal analysis. We acknowledge that the meanings of data altruism and related terms include those found beyond academic literature.

Our analysis of current definitions and understandings of data altruism reveals frequent comparisons with data donation, data philanthropy, data crowdsourcing and data solidarity. To better understand how data altruism relates to these related terms, we analyzed the salient similarities and differences between the concepts. We learn, firstly, that data altruism can be understood narrowly as data donation activities in the sense of data donation or data philanthropy, which includes all activities related to selfless generation, collection, and processing of data in the public interest. This understanding promotes the understanding of data altruism as ascribing voluntary data sharing with a certain gift logic and data handling practices as, arguably, civic work. Secondly, like data crowdsourcing in the US, the data altruism regulatory regime includes institutional work to regulate, validate, and standardize voluntary data sharing for objectives in general interest (see Fig. 1). Thirdly, data altruism is a mode of data governance like data solidarity, which calls attention to the distribution of potential harms and benefits along the entire altruistic data value chain.

Our analysis of actors and roles reveals and illustrates a variety of players envisioned to participate in data altruism. Most notable is the new legal entity, RDAO, which will be allowed exclusively to communicate using the novel EU data altruism common logo, which emerges as a recognizable symbol of legitimacy, aiming to foster trust and credibility among (potential) data altruism actors. Furthermore, we find that especially individuals are expected to be empowered to share their personal data through new consent mechanisms and specialized SSI-based digital tools. At the same time, we learn that the role of technology providers and the potential dependence of other actors on them have received little attention in the literature so far.

Additionally, our analysis of data expected to be voluntarily shared and the potential purposes strongly suggests that the data altruism value chain proposition is largely promissory, as reflected in the dearth of empirical work, and limited and general examples of specified objectives of general interest found in the academic literature. There is a need to build more empirical knowledge about how specific types of data are associated with specific application purposes, and how we can assess the outcomes of data sharing and use.

The review has certain limitations associated with coverage as it focuses narrowly on the altruistic dimension of voluntary data sharing. As such, it does not fully capture the consequences of data sharing, other than the purported objectives of general interest. Furthermore, the in-depth sample is small due to the emergent nature of the field. We limited our search to academic articles to maintain focus on the contested novelty claims found in the academic literature. The implementation of ante litteram data altruism projects are addressed in few studies. Additionally, it is often challenging to distinguish the promises of data altruism from factual statements about effects of data sharing [17, 19]. Overall, the academic research on data altruism is still emerging and optimistic, with few explicit mentions of potential challenges or risks.

6 Conclusion and Suggestions for Future Research

Thus, how can data altruism be understood as data donation? Answering the research question based on our findings, while acknowledging the methodological limitations, warrants prudence. Data altruism can be understood as the introduction of a formal 'data donation' regime regulated by the EU. Data altruism represents a deliberate effort to institutionalize innovative, non-commercial voluntary data sharing practices aligned with EU policy objectives or *objectives of general interest*. In this sense, we urge future scholars to produce knowledge sensitive to its sociopolitical context. For instance, the European data altruism consent form suggests a distinctively contractarian view on data altruism, which legitimates data sharing based on the explicit consent or permission of the donor. This is in stark contrast to our insights gleaned from the data solidarity literature, wherein the collective interest may prevail over individual autonomy in the case of low-risk data sharing.

Our clarified understanding of the distinctiveness of data altruism lays the groundwork for further research on data governance in the digital government domain and beyond. Our conceptual refinement hints at the important roles played by diverse actors, including governmental, across various fields in shaping and legitimizing data altruism norms, standards, and policies. Therefore, we encourage future research to not only study data altruism as a digital and regulatory, but also an evidently social, innovation. We encourage future research to explore how voluntary registration in a public national register of RDAOs can increase trust in data sharing [44], to examine the interactions between existing national data donation regimes (e.g., France, Denmark, and Germany) and the EU data altruism regime, and to publish descriptive-analytical studies on the bottom-up and top-down initiatives in this new regulatory space. Relevant future research should also consider who benefits from the exclusive use of the data altruism common logo and who may be excluded or disadvantaged by its adoption. Additionally, how can RDAOs address and mitigate skepticism or resistance from stakeholders, including concerns about privacy and data ownership? By exploring the interactions between technological advancements, regulatory frameworks, and sociopolitical dynamics, future research can contribute to a useful understanding of how data altruism can achieve its objectives of general interest. This approach will enable policymakers to identify valuable opportunities for collaboration, provide a scientific evidence base for the data altruism rulebook currently in the making and, ultimately, to enable data altruism to deliver on its promissory benefits.

Acknowledgments. We gratefully acknowledge the valuable support and insightful feedback provided by Flor Avelino, Bonno Pel, and Koen Frenken on earlier versions of this manuscript. We also thank the anonymous reviewers for their useful suggestions.

Disclosure of Interests. The authors have no competing interests to declare that are relevant to the content of this article.

References

1. Mazur, J., Słok-Wódkowska, M.: Access to information and data in international law. Nord. J. Int. Law 310–338 (2022). https://doi.org/10.1163/15718107-91020004
2. Green, B.: Data science as political action: grounding data science in a politics of justice, pp. 249–265 (2021). https://doi.org/10.23919/JSC.2021.0029
3. Alemanno, A.: Big data for good: unlocking privately-held data to the benefit of the many. Eur. J. Risk Regul. **9**(2), 183–191 (2018). https://doi.org/10.1017/err.2018.34
4. McNutt, J.G., Goldkind, L.: Civic technology and data for good: evolutionary developments or disruptive change in E-participation?. In: Research Anthology on Citizen Engagement and Activism for Social Change, pp. 1330–1345. IGI Global (2022)
5. Lamchek, J.S.: Ensuring data science and its applications benefit humanity: data monetization and the right to science. Hum. Rights Law Rev. **23**(3), ngad018 (2023). https://doi.org/10.1093/hrlr/ngad018
6. Lalova-Spinks, T., Meszaros, J., Huys, I.: The application of data altruism in clinical research through empirical and legal analysis lenses. Front. Med. **10**, 1141685 (2023). https://doi.org/10.3389/fmed.2023.1141685
7. Ferrè, G.R.: Data donation and data altruism to face algorithmic bias for an inclusive digital healthcare. BioLaw J. – Riv. BioDiritto (1), 115–129 (2023)
8. Templier, M., Paré, G.: A framework for guiding and evaluating literature reviews. Commun. Assoc. Inf. Syst. **37**, 112–137 (2015). https://doi.org/10.17705/1CAIS.03706
9. Torraco, R.J.: Writing integrative literature reviews: guidelines and examples. Hum. Resour. Dev. Rev. **4**(3), 356–367 (2005). https://doi.org/10.1177/1534484305278283
10. Torraco, R.J.: Writing integrative literature reviews: using the past and present to explore the future. Hum. Resour. Dev. Rev. **15**(4), 404–428 (2016). https://doi.org/10.1177/1534484316671606
11. Wolfswinkel, J.F., Furtmueller, E., Wilderom, C.P.M.: Using grounded theory as a method for rigorously reviewing literature. Eur. J. Inf. Syst. **22**(1), 45–55 (2013). https://doi.org/10.1057/ejis.2011.51
12. Gusenbauer, M., Haddaway, N.R.: Which academic search systems are suitable for systematic reviews or meta-analyses? Evaluating retrieval qualities of Google Scholar, PubMed, and 26 other resources. Res. Synth. Methods **11**(2), 181–217 (2020). https://doi.org/10.1002/jrsm.1378
13. Zichichi, M., Ferretti, S., D'Angelo, G., Rodríguez-Doncel, V.: Data governance through a multi-DLT architecture in view of the GDPR. Clust. Comput. **25**(6), 4515–4542 (2022). https://doi.org/10.1007/s10586-022-03691-3
14. Mirval, J., Bouganim, L., Sandu-Popa, I.: Practical fully-decentralized secure aggregation for personal data management systems. In: 33rd International Conference on Scientific and Statistical Database Management, Tampa, FL, USA, , pp. 259–264. ACM (2021). https://doi.org/10.1145/3468791.3468821
15. Esteves, B., Rodríguez-Doncel, V.: Semantifying the governance of data in Europe. In: CEUR Workshop Proceedings, vol. 3235 (2022)
16. Jasserand, C.: Research, the GDPR, and mega biometric training datasets: opening the pandora box. In: 2022 International Conference of the Biometrics Special Interest Group (BIOSIG), Darmstadt, Germany, pp. 1–6. IEEE (2022). https://doi.org/10.1109/BIOSIG55365.2022.9897040
17. Berti Suman, A., Balestrini, M., Haklay, M., Schade, S.: When concerned people produce environmental information: a need to re-think existing legal frameworks and governance models?. Citiz. Sci. Theory Pract. **8**(1), 10 (2023). https://doi.org/10.5334/cstp.496

18. Kruesz, C., Zopf, F.: European union • the concept of data altruism of the draft DGA and the GDPR: inconsistencies and why a regulatory sandbox model may facilitate data sharing in the EU. Eur. Data Prot. Law Rev. **7**(4), 569–579 (2021). https://doi.org/10.21552/edpl/2021/4/13
19. Berti Suman, A., Heyen, N.B., Micheli, M.: Reimagining health services provision for neglected groups: the 'personalization from below' phenomenon. Front. Sociol. **8**, 1052215 (2023). https://doi.org/10.3389/fsoc.2023.1052215
20. Susha, I., Grönlund, Å., Van Tulder, R.: Data driven social partnerships: exploring an emergent trend in search of research challenges and questions. Gov. Inf. Q. **36**(1), 112–128 (2019). https://doi.org/10.1016/j.giq.2018.11.002
21. Prainsack, B.: Data donation: how to resist the iLeviathan. Ethics Med. Data Donation 9–22 (2019). https://doi.org/10.1007/978-3-030-04363-6_2
22. Hummel, P., Braun, M., Dabrock, P.: Data donations as exercises of sovereignty. Ethics Med. Data Donation 23–54 (2019). https://doi.org/10.1007/978-3-030-04363-6_3
23. Skatova, A., Goulding, J.: Psychology of personal data donation. PLoS ONE **14**(11), e0224240 (2019). https://doi.org/10.1371/journal.pone.0224240
24. Taddeo, M.: Data philanthropy and the design of the infraethics for information societies. Philos. Trans. R. Soc. Math. Phys. Eng. Sci. **374**(2083), 20160113 (2016). https://doi.org/10.1098/rsta.2016.0113
25. Taddeo, M.: Data philanthropy and individual rights. Minds Mach. **27**(1), 1–5 (2017). https://doi.org/10.1007/s11023-017-9429-2
26. George, J., Yan, J.K., Leidner, D.: Data philanthropy: an explorative study (2019). https://scholarspace.manoa.hawaii.edu/server/api/core/bitstreams/aab6eab6-bec1-47c7-a8fe-ade0c3f10a7f/content
27. Ajana, B.: Communal self-tracking: data philanthropy, solidarity and privacy. Self-Track. Empir. Philos. Investig. 125–141 (2018). https://doi.org/10.1007/978-3-319-65379-2_9
28. McKeever, B., Greene, S., MacDonald, G., Tatian, P.A., Jones, D.: Data philanthropy: unlocking the power of private data for public good (2018). https://policycommons.net/artifacts/631042/data-philanthropy/1612321/. Accessed 13 Feb 2023
29. Hillebrand, K., Hornuf, L., Müller, B., Vrankar, D.: The social dilemma of big data: donating personal data to promote social welfare. Inf. Organ. **33**(1), 100452 (2023). https://doi.org/10.1016/j.infoandorg.2023.100452
30. Awasthi, P., George, J.: Harmonizing strategic advantage with social good through data philanthropy. In: GlobDev 2019 (2019). https://aisel.aisnet.org/globdev2019/5
31. Murphy, R.J., Parsons, J.: Where does the data go? Data modelling and reuse in crowdsourcing for social innovation, vol. 14s (2020). https://aisel.aisnet.org/icis2020/digital_innovation/digital_innovation/14/
32. Choy, K., Schlagwein, D.: Crowdsourcing for a better world: on the relation between IT affordances and donor motivations in charitable crowdfunding. Inf. Technol. People **29**(1), 221–247 (2016). https://doi.org/10.1108/ITP-09-2014-0215
33. Salido-Andres, N., Rey-Garcia, M., Alvarez-Gonzalez, L.I., Vazquez-Casielles, R.: Mapping the field of donation-based crowdfunding for charitable causes: systematic review and conceptual framework. Volunt. Int. J. Volunt. Nonprofit Organ. **32**(2), 288–302 (2021). https://doi.org/10.1007/s11266-020-00213-w
34. Goncalves, J., et al.: Crowdsourcing on the spot: altruistic use of public displays, feasibility, performance, and behaviours. In: Proceedings of the 2013 ACM International Joint Conference on Pervasive and Ubiquitous Computing, Zurich, Switzerland, pp. 753–762. ACM (2013). https://doi.org/10.1145/2493432.2493481
35. Baruch, A., May, A., Yu, D.: The motivations, enablers and barriers for voluntary participation in an online crowdsourcing platform. Comput. Hum. Behav. **64**, 923–931 (2016). https://doi.org/10.1016/j.chb.2016.07.039

36. Bunz, M., Vrikki, P.: From Big to democratic data: why the rise of AI needs data solidarity. In: Democratic Frontiers. Taylor & Francis (2022)
37. Prainsack, B., Buyx, A.: A solidarity-based approach to the governance of research biobanks. Med. Law Rev. **21**(1), 71–91 (2013). https://doi.org/10.1093/medlaw/fws040
38. Braun, M., Hummel, P.: Data justice and data solidarity. Patterns **3**(3), 100427 (2022). https://doi.org/10.1016/j.patter.2021.100427
39. Prainsack, B., Buyx, A.: Thinking ethical and regulatory frameworks in medicine from the perspective of solidarity on both sides of the Atlantic. Theor. Med. Bioeth. **37**(6), 489–501 (2016). https://doi.org/10.1007/s11017-016-9390-8
40. PorsdamMann, S., Savulescu, J., Sahakian, B.J.: Facilitating the ethical use of health data for the benefit of society: electronic health records, consent and the duty of easy rescue. Philos. Trans. R. Soc. Math. Phys. Eng. Sci. **374**(2083), 20160130 (2016). https://doi.org/10.1098/rsta.2016.0130
41. Jones, K.H., Laurie, G., Stevens, L., Dobbs, C., Ford, D.V., Lea, N.: The other side of the coin: harm due to the non-use of health-related data. Int. J. Med. Inf. **97**, 43–51 (2017). https://doi.org/10.1016/j.ijmedinf.2016.09.010
42. Bak, M.A.R., Ploem, M.C., Tan, H.L., Blom, M.T., Willems, D.L.: Towards trust-based governance of health data research. Med. Health Care Philos. **26**(2), 185–200 (2023). https://doi.org/10.1007/s11019-022-10134-8
43. De Franceschi, A.: Personal data as counter-performance. In: Senigaglia, R., Irti, C., Bernes, A. (eds.) Privacy and Data Protection in Software Services. Services and Business Process Reengineering. Springer, Singapore, pp. 59–71 (2022). https://doi.org/10.1007/978-981-16-3049-1_6
44. Micheli, M., Farrell, E., Carballa, S.B., Posada, S.M., Signorelli, S., Vespe, M.: Mapping the landscape of data intermediaries. JRC Publications Repository. https://publications.jrc.ec.europa.eu/repository/handle/JRC133988. Accessed 05 Jan 2024
45. Zuboff, S.: Surveillance capitalism and the challenge of collective action. New Labor Forum **28**(1), 10–29 (2019). https://doi.org/10.1177/1095796018819461
46. Data Governance Act explained | Shaping Europe's digital future. https://digital-strategy.ec.europa.eu/en/policies/data-governance-act-explained. Accessed 05 Mar 2024
47. Curry, E., et al.: The European big data value ecosystem. Elem. Big Data Value **3**, 3–19 (2021). https://doi.org/10.1007/978-3-030-68176-0_1
48. Cavanillas, J.M., Curry, E., Wahlster, W. (eds.) New Horizons for a Data-Driven Economy. Springer, Cham (2016). https://doi.org/10.1007/978-3-319-21569-3
49. Sagiroglu, S., Sinanc, D.: Big data: a review. In: 2013 International Conference on Collaboration Technologies and Systems (CTS), San Diego, CA, USA, pp. 42–47. IEEE (2013). https://doi.org/10.1109/CTS.2013.6567202
50. Taylor, L.: What is data justice? The case for connecting digital rights and freedoms globally. Big Data Soc. **4**(2), 2053951717736335 (2017). https://doi.org/10.1177/2053951717736335
51. Montuori, A.: Literature review as creative inquiry: reframing scholarship as a creative process. J. Transform. Educ. **3**(4), 374–393 (2005). https://doi.org/10.1177/1541344605279381

A Method for the Collaborative and Semi-automated Generation of Conceptual Models from Legal Regulations in Public Organizations

Binh An Patrick Nguyen[1]( ) and Hendrik Scholta[2]

[1] University of Münster, 48149 Münster, Germany
patrick.nguyen@ercis.uni-muenster.de
[2] German University of Administrative Sciences Speyer, 67346 Speyer, Germany

Abstract. Legal regulations and conceptual models are important for public organizations. Conceptual models are means for complexity reduction in the design and customization of information systems. Legal regulations are important for public organizations as they specify the services that the organizations offer to citizens and businesses. However, the operationalization of legal regulations is challenging because they leave room for interpretation and there is a high number of involved actors. These actors comprise of domain experts and IT experts within a public organization, but also other public organizations on the same or different levels of government. In consequence, there are many actors involved in the operationalization and execution of legal regulations for public services who would benefit from the use of conceptual models. To provide support for these actors, we address the following research goal: Design of a method for the collaborative and semi-automated generation of conceptual model from legal regulations in public organizations. In the course of our design science research approach, we derived requirements for the solution based on interviews with thirteen public officials. We conceptually developed the method and evaluated it with an illustrative scenario and expert interviews with six public officials. The evaluations reveal the general potential usefulness and intended use of our method.

Keywords: legal modelling · law modelling · digital government · conceptual modelling · public administration · design science research

1 Introduction

Legal regulations (LRs) require citizens to perform certain actions, such as register their residence, register their vehicles, and pay taxes [1]. To do so, citizens require a point of contact where they can perform those actions, usually by handing over information relevant regarding their cases. Public organizations (POs) are these points of contact and enable citizens to fulfil their duties lawfully by offering public services to them [2]. To enable citizens to be lawful, also these public services must be lawful and comply with LRs.

However, it is not trivial for POs to align their services to LRs because public officials' interpretations of LRs are not unique and different officials can interpret LRs differently. LRs are often formulated in a way that few words cover a high number of possible scenarios, making it difficult to determine the application of LRs to cases [3]. POs do not only need to link their services and activities to LRs, but also interpret them, usually with the assistance of additional regulations set forth by higher-level POs. While there is generally a chain from national to regional to local POs, due to political influence on LRs and their local implementation, the interpretations can vary even after guidance by POs located at the national and regional levels [4].

To reduce different interpretations and applications of LRs between people, having a common ground and shared understanding is important for the execution of LRs [5]. Conceptual models (CM) such as process models and data models can depict recommended operationalizations of LRs and facilitate the communication between experts of the domain and experts in IT [6]. Each of the experts contributes their understanding in observing an object of interest and creating a model which represents the object's relevant parts. The model is created in a dialogue between both experts checking each other's conceptualization of the object [6].

However, a LR has more stakeholders than a domain expert and an IT expert [7]. Apart from the intra-PO collaboration between domain and IT expert, there are vertical and horizontal inter-PO collaborations possible or taking place. Horizontal collaborations between POs on the same level are beneficial to share CMs between POs and thereby create a common understanding of a LR across them. Such exchanges reduce the necessity to analyse a LR many times in different POs and ideally require an analysis for many POs only once. Additionally, LRs may involve POs vertically, as national governmental units may give additional instructions for POs on lower levels. A LR can be passed by a national or regional parliament, while local POs execute it.

In consequence, there are many actors involved in the specification, interpretation, operationalization and execution of LRs in POs which would benefit from the use of CMs [8]. To support such a collaboration, a semi-automated method for the creation of CMs is useful to adequately bring together the different actors and release them from avoidable manual and redundant work. Hence, we set our research goal as follows: *Design of a method for the collaborative and semi-automated generation of CMs from LRs in POs.* We addressed the research goal following the design science research paradigm [9]. We derived requirements for the method (Sect. 4.1), developed the method's design (Sect. 4.2) and conducted an evaluation with an illustrative scenario and interviews with practitioners (Sect. 4.3).

2 Research Background

2.1 Conceptual Models

CMs are imposed by external entities upon an object of interest to represent relevant parts of it [10]. There are various types of conceptual models that help to design information systems because they reduce the complexity of reality [10]. For example, there are process models that depict only the processes and their activities that employees execute in an organization [11], data models that represent the structure of a database or organizational charts that visualize organizational units and their relationships.

A model is created in a dialogue between a target domain expert and a IT expert [6]. The dialogue serves in discussing problems because a IT expert or "modeler may observe the universe directly, but still depends on [...] representations [the domain expert] brings forth to get (more) accurate information" [6, p. 131]. This is especially relevant in law execution, as a law should be executed the same in every part of its area of validity. Having a common model incorporating knowledge from both sides – the domain expert and the IT expert – can assist in reaching consensus about the interpretation of the legal concepts [12].

2.2 Legal Information Systems

The main purpose of legal information systems is to make the high amount of LRs accessible and understandable for legal experts and non-legal experts [13, 14]. Examples include displaying only LRs that are relevant to a certain case or showing cases similar to the one at hand [15].

As potential idea for future research in legal information systems, LRs could be tested if they can be operationalised as intended in controlled environments [16]. Additional information can put legal concepts into perspective for non-legal experts to understand the requirements for digitalising LRs [17]. Legal information systems can assist in reducing both the cause and effect of redundant definitions [18, 19].

As part of legal information systems, the literature proposed methods that transform LRs into CMs [e.g., 20–29]. We conducted a literature review on such methods for the creation of CMs from LRs and found 40 of them [30]. The majority of those methods was published over 10 years ago, indicating a need to refresh knowledge and for additional input from different domains. Outputs of such methods are CMs like data models, process models, business rules, or legal ontologies [30]. Most of the methods take unstructured text as input before refining it into a structured text in the form of XML templates or specific variants of XML like Akoma Ntoso or LegalDocML [28]. This is to ensure that the XML variant it translates to is usable by the method itself. It is not guaranteed that the transformation of one method can be used in another method [30].

Methods like the POWER method [29] or VLPM [22] create models from LR text, however those are often either only nation-level models or finished unmodifiable models. Furthermore, their models focus only on one LR at hand to specify explicitly stated concepts and their relationships.

Methods like SPGM take local specifics into consideration, by having public officials of various levels modelling processes collaboratively, however there is no systematic step intended to inject and track the local changes to the national model or to make it configurable [31]. Essentially, there is a lack of upwards communication of changes and decisions [30].

Additionally, the methods differ based on which development phase an LR is in. A LR in draft is not legally binding yet, but POs can use it to check their processes and prepare for potential changes [30].

Although there are numerous methods that generate CMs from LRs in POs, they do not account for the collaborative nature within a PO and vertically and horizontally across POs that is necessary for an execution-focused generation of CM.

3 Research Approach

Design Science Research (DSR) proposes to include design-oriented methodologies from creative and manufacturing fields like engineering for the design of information systems [9]. Hevner et al. [9] differentiate four output types of DSR projects: constructs, instantiations, methods, and models. In this research, we create a method to conceptualize a process to create models from legal text. The DSR process can be separated into six steps [32]. In this research, we performed the steps as follows:

- Phase 1 - *Motivation and problem identification*: Based on the need in practice and existing literature about generating CMs from textual LRs (Sect. 2.1), we identified the gap that existing methods do not account for the multiple actors who should work together collaboratively in the generation of CMs from LRs.
- Phase 2 - *Objectives of a solution*: We conducted qualitative interviews with public officials in Germany (ten interviews with thirteen public officials from digitalisation departments of eight cities and a municipal service provider) to further contextualize the gap. All interviewees are involved in the operationalization of LRs for public services. The interview guide consisted of the sections "introducing the interviewee" where the interviewees told us about their role and experience in their POs, "the current state of operationalising LRs" where they told us about how they currently operationalise LRs, and "the ideal state of operationalising LRs" where they told us how CMs could be helpful for the future of operationalisation. The interviews were transcribed, and we analysed them qualitatively to obtain requirements for our solution.
- Phase 3 – *Design and development of an artefact:* Based on the requirements, we developed a method to support the different actors in operationalizing a LR from text to model. As a de facto industry standard, BPMN 2.0 was used to model the method's processes.
- Phases 4 and 5 – *Demonstration* and *Evaluation*: We evaluated the method through an illustrative scenario and expert interviews [33]. For the illustrative scenario, we used an exemplary law, the German Bundesreisekostengesetz [English: federal traveling expenses law], and applied components of the method to generate models from the law. The guide for the expert interviews consisted first of an introduction followed by a shortened ideal state block similarly to the interviews of phase 2. In the next section, the interviewees evaluated the general idea of the method before being presented with mock-ups running through parts of the Bundesreisekostengesetz to demonstrate the functionality of the method in detail. The interviewees were six public officials from digitalization departments with experience ranging from freshly started to 28 years in the PO. The experts were from different sub areas of digitalisation in their respective organizations, the roles ranging from overall IT department heads over specific digitalisation officers to regional coordinators. Their backgrounds were diverse, as they had studied law, public administration, business administration, or computer science.

The interviews were again transcribed and analysed qualitatively. We integrated the evaluation results into a new version of the method's requirements and design.
- Phase 6 - *Communication*: This paper and succeeding papers are part of the project's communication. Additionally, as a DFG-funded research project, an intermediate and a final report will be written to communicate results and the status of the project.

4 Results

4.1 Requirements

Busch [34] proposes a process that public officials execute to deliver a public service and that consists of the following steps: Decision need occurs, collect information, identify decision alternatives, weigh evidence, choose decision alternative, evaluate, and implement action. Based on this process, we derive *process models, forms*, and *decision trees* as central artefacts for public services and as models to be generated by our method. Process models are relevant to detail Busch's [34] generic process for specific services. Additionally, collecting information is typically done with paper-based or digital forms detailing the information to be asked for. Hence, forms can be represented as CMs, both to maintain a common standard and as a data structure for legal information systems. Finally, public officials need to come to a decision regarding a case. To support decisions and visualize the rules that underly them, decision trees can point out both the decisions to be made and the facts required to make an informed decision.

Based on the selection of process models, forms, and decision trees as the models to be generated with our method, we derived the following requirements (REQs):

REQ1 - Handle Different User Groups and Stakeholders: The ability to trace both the source or a LR and the jurisdiction of a user is important as they determine a user's authority to make a decision regarding a LR. It was suggested by multiple interviewees that they would take this authority into consideration when determining whether to use an existing text annotation of an LR. The jurisdiction would lend credibility to a user based on the level (a national level annotator would receive more credibility than a local level annotator regarding a LR from the national level).

REQ2 - Provide and Handle Markups and Annotations: LRs need to be annotated to indicate what information a text provides (e.g., form field). This can be done manually or with automated support through artificial intelligence. We differentiate between markups, i.e., the marked words of an annotation, and the attributes (metadata) of an annotation, e.g., the name of a process activity that is to be generated from an annotation. Our illustrative scenario revealed that an annotation can contain marked words that may not be sequenced directly next to each other in the text (e.g., "any costs incurred [...] will be reimbursed"). Additionally, each attribute can belong to multiple markups (e.g., the attribute "Process activity: Reimburse costs" can be linked to various text passages). A change to a shared attribute can cascade to all linked markups.

REQ3 - Create New Models of Different Model Types: The method must be able to derive models from text. Those models are also to be integrated to link models together (e.g., a decision tree element being the condition of a XOR gateway in a process model). To track and trace model elements, they need to be further described by attributes like legal source. Moreover, various LRs can lead to similar CMs. Therefore, once a model has been created for one of the LRs, the model can be reused and slightly adapted for the other LRs. It is not necessary to create their models from scratch. Finally, users should be able to adapt the created general models to individual specifics within given legal frames.

REQ4 - Provide and Handle Already Existing Models: POs typically already have collections of models that depict current practices of these organizations (e.g., process models of the activities currently executed in service delivery). Therefore, it is interesting for POs to reveal whether their current practices comply with LRs and which of the activities are required by law and which are executed because of other reasons. To allow such comparisons, models need to be processed in frequently used formats.

REQ5 – Import and Handle Legal Texts: Texts need to be imported into the database. While doing that, common checks like duplicates, missing references or updated references need to be included. This in combination with version control allows for technical support of legal compliance. Changes of LRs should be highlighted and annotations of unchanged text passages should be transferred from earlier versions to newer versions, which can reduce overhead as only new annotations in the changed areas are required. Multiple services can be derived from one text, which means that in addition to an attribute for models and model elements to trace their origin, texts can have attributes to track the services they belong to.

4.2 Artefact

We first present the overview of the method (See Fig. 1). The user starts the process manually and selects an LR to be the basis for new CMs of public services. The texts are then annotated with model-related information. This includes both a process management view on how to realize the service and a legal view on what the relevant objects are and what constraints are set forth by an LR. In case that different users do not agree in their annotations, these conflicts are resolved before three model types are realized for the LR. The CMs are then integrated with each other, e.g., a process flow gateway can correspond to a decision node or chance node in a decision tree. Finally, the user

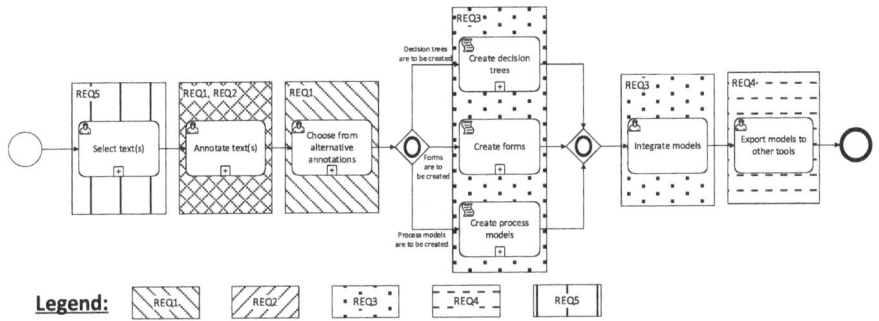

Fig. 1. Top-level process of the method.

exports the CMs for use and refinement in other systems such as process modelling tools or forms management systems. In the following, we detail each of the method's steps.

Select Text(s). REQ5 deals with the import and handling of legal texts (See Fig. 2). As a basis, a directory of all texts available is shown to a user. A user selects a text and if the text is not in the database, then it needs to be imported by the system automatically.

The text is then assigned to a legal domain to match legal experts, which may vary depending on the government structure and levels (e.g., ministry of finance and local tax office for tax LRs). Based on jurisdiction, the appropriate expert or collaboration partner can be notified about an imported text (e.g., a domain expert specialized in finance is notified if a process manager as IT expert imports a LR related to taxes).

Subsequently, the system checks for duplicates of the text to reduce work. This is relevant as LRs may change over time, but depending on the situation, prior versions may still need to be accessible. For instance, a citizen received a service at the time of an old LR and the LR changed later on. Hence, the model derived from the old LR is still required for traceability reasons.

Additionally, texts are pre-processed in this step. This includes highlighting text changes compared to previous versions of a text to reduce redundant work on areas already annotated.

The user can use a recommender system to draw upon prior annotations and use them as a way to keep common definitions and terms standardized. For this purpose, the database is checked for existing annotations, which can then be transferred to the next text.

LRs are also linked through the references in the text and the links are then manually checked for correctness. It may be required to check all documents for references as, for example, German LR references are not always bi-directional (both documents referring to each other), but rather unidirectional (only one document refers to the other).

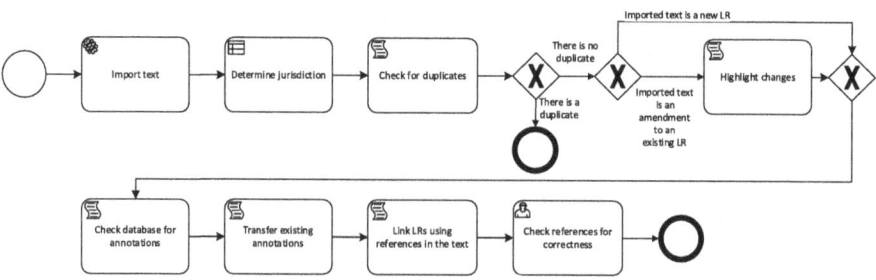

Fig. 2. Select text(s).

Annotate Text(s). As REQ2 requires the method to enable the annotation of text, we propose the ability to markup words (See Fig. 3). At the beginning, all existing annotations of all texts are retrieved from the database to serve as an index and recommender function for annotations. A user can mark additional words to extend an existing annotation, since we want to enable the adaption of annotations. This is done for all three model types.

A prerequisite to enable REQ3 later on is fulfilled by the possibility to use one single markup for multiple model types as they are not mutually exclusive. For example, decision trees and process models can be combined where the process models depict different branches and the decision trees depict the ways on how to choose one of the alternatives. With the markup in the LR, a model created from an annotation can track its source by setting the appropriate attribute.

REQ1 puts a focus on collaborative work. Users discuss annotations to create a common ground of understanding, both on the validity and suitability of an annotation and appropriate resolution if both sides do not agree on an annotation. The resolution comprises a change to the annotation such that all involved users can agree to it. To provide creditably of the involved users, it is also of importance to track the users' jurisdiction and the LRs' legal domain. When the final annotation is approved, it is released to the database from where it can be retrieved to create a CM or be used, for example, by a recommender system for future texts.

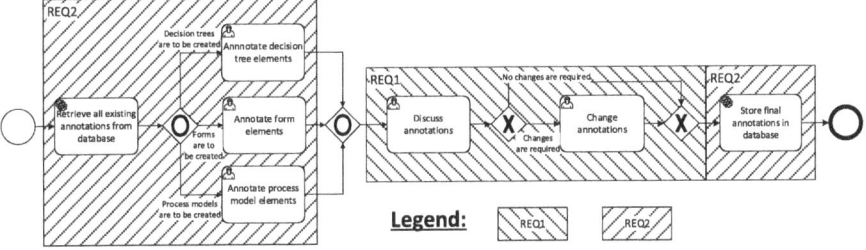

Fig. 3. Annotate text(s).

Choose from Alternative Annotations. To address REQ1 and account for the different people creating annotations for their services, the user has now to decide which of various valid annotations is suitable for their own specific service (See Fig. 4). Such situations can occur if various users annotate LRs without discussing their annotations. Exemplarily, we assume two annotations for the field groups "person" and "person of public interest". The latter one details the former and has a higher granularity. The user should be presented with both valid annotations and decide which annotation matches their purposes best.

REQ2 is addressed by differentiating annotations that may concern the same part of an LR. This can be done by the use of additional attributes and the combination thereof like user, timestamp, field of application or service.

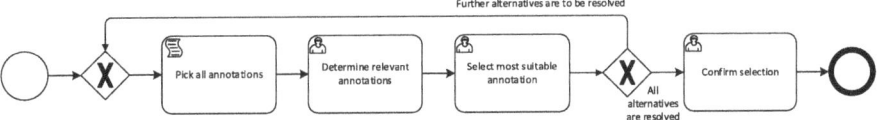

Fig. 4. Choose from alternative annotations.

Create Models. REQ3 requires the generation of CMs and the method separates the modelling process by the three model types (See Fig. 5). Each of them has a slightly different process. In general, model fragments are created in an automated manner from annotations with the source as an attribute. Those individual fragments are checked collaboratively for legal compliance before being linked to larger models.

Process models are created in two ways: (1) Transforming the annotations into new process fragments or (2) detecting similar annotations in other LRs that have already been transformed into process fragments and suggesting these fragments for the annotation at hand. In the first case, the user can link fragments to create bigger fragments while in the latter case the user needs to check the suitability of the suggested fragment.

Decision trees are created differently. First, the user analyses the annotations whether they represent decisions in the responsibility of the decision-maker (decision node), uncertain events not in the responsibility of the decision-maker (chance node) or consequences (end node). After these sets have been discovered, intersections of them are checked to see if consequences can lead to further decisions or events in other decision trees.

Forms are identified by title and contain fields and field groups. Based on the annotations, fields and field groups are identified. Using both annotations and the database of forms, fields can be assigned to field groups. However, in case of a conflicting assignment (e.g., unnecessary fields or fields missing/wrongly assigned), a user has to resolve it before approving the model.

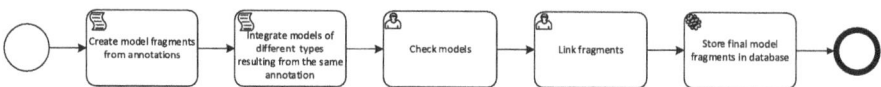

Fig. 5. Create models.

Integrate Models. Since each model type is created independently from the other types, we need to integrate the resulting models to fulfil REQ3. This integration of all models for an intended public service is useful to ensure consistency of elements across models of different types originating from the same annotation. An example would be the representation of a decision node or chance node as a XOR gateway or linking a form as a data object to a process activity.

Export Models to Other Tools. Finally, to complete REQ4 and to allow the modification of created models, these can be exported into modelling tools for editing purposes. Furthermore, the exported models can be compared to existing models for either compliance checks through, for instance, process model matching approaches [35]. Alternatively, the newly created models can be merged into existing models to increase their compliance. The exports also include the model's legal source for traceability.

4.3 Evaluation

Illustrative Scenario. We evaluated our method using an illustrative scenario. In general, the scenario showed that the major steps work properly. In response to the results of the evaluation, we made the following adaptions to the requirements and design of the method.

First, we realized that decisions are often specified with only one consequence, resulting in incomplete decision trees. If we wanted to complete such decision trees, we would need to incorporate model elements that are not specified by LRs. Sometimes we recognized that it is not even clear in which order the elements could be connected at all. Since we want to provide only CMs that are completely determined by LRs, we changed REQ3 to generate model fragments instead of complete CMs.

Second, the illustrative scenario showed that an annotation might consist of words spread across the text. Therefore, REQ2 requires annotations to consist of separated words.

Third, we experienced that because relevant information regarding the LRs can have different levels of granularity and this leaves a lot of room for interpretation, it is difficult to create one single coherent model from these texts. This was especially crucial if a single LR defines multiple services. Therefore, to account for this aspect, we adapted REQ2 to allow for individual attributes of annotations such as the name of the service the annotation relates to.

Expert Interviews. After enhancing the method with the insights from the illustrative scenario, a further evaluation was conducted with expert interviews. Overall, the evaluation was positive. Interviewees emphasized the novelty and necessity of creating legally compliant models.

Some interviewees saw the tool as a collaboration enabler, where IT experts from digitalisation departments and legal experts from domain departments can use it to ensure a common ground through CMs. One interviewee stated that this would help not *"being overwhelmed with legalese for 60 min"*, as the models would translate the concepts the domain experts know into a language the IT experts can understand. One interviewee's first reaction was enthusiastic: *"When can I use this?"*. This is seconded by another's *"Why does this not exist yet?"*. Additionally, the groundwork laid by the created model fragments would help in reducing change requests as they *"build a base template for collaboration which truly is rooted in legal norms"*.

There are multiple tools for POs to create forms, model processes, or in general support them in their daily operation: *"It is important to me to be able to export this [the models], because [Tool A] works only with [Tool B] and that is already difficult."* Therefore, it is necessary that the method allows to export the models in frequently used formats (see REQ4).

Another interviewee stressed the importance of different user roles because *"There are pay grades, specifically so one knows who is authorized to make decisions, or how far your authority goes. Yet everything is to be controlled again"*. The method we present can assist in structuring and alleviating the fear of doing something wrong by making sure annotations can be controlled and double-checked. Especially the roles would allow to check if different expertise was applied in creating the models for a specific service: *"ideally [LRs] should not exclusively be designed by legal experts but also by practice-oriented people or those who actually execute them locally."* (see REQ1).

An intensively discussed question was *"who should use it?"*. Within a PO, the application of the method ranges from digitalisation experts using it as a leverage against legal experts for indicating activities and form fields that are not legally required and impose additional burden on citizens, to domain experts discussing among each other or in collaboration with digitalisation experts. Also, the legislation could use this tool to centrally create models to hand down to the local POs.

Overall, a major concern was that the threshold between actual benefit and administrative overheads like logins and complexity could be too high for a successful adoption: *"Often a tool is opened, people read a moment and don't know what to do. Then they close it and think – if I have time I'll deal with that – and push it away."* Hence, the use should also be guided well.

One interviewee pointed out that the database of annotations should be used on a national level to ensure standards, or at least being able to see and copy processes of other POs and adapt them for own use cases. Additionally, the interviewee was adamant that the method should be used to network different POs to access the *"power of many"* and thus discovering best practices for implementing a LR. This sentiment was echoed by several interviewees, as they stated broadly that an integrated database would reduce the *"wild growth"* of individual variants of the same legally restricted process. This depends on the complexity of the process and service to be modelled. *"If I can model it within five minutes, I won't spend 10 min searching through other administrations."* This aspect was integrated into REQ4 such that frequently used formats enable the comparison of created models to existing models.

A concern regarding who would supply the models and who would use them was also raised by the interviewees: *"[metropolitan city X] has a different staffing ratio than [midsized city A]. That X will be more of a giver than a taker at this point, that will be the case. [...] We would love to give as well, especially on the lower levels of administrations we split up a process and everyone supplies a small part of the greater [service]."*

From their perspective as a process manager, one interviewee pointed out that changes in LRs and thus annotations need to cascade through all models. Especially the rooting of the model fragments in LRs was seen as a positive aspect by the majority of the interviewees: *"thus it is truly connected to the law and one knows where it comes from. So basically, I don't make it up out of thin air, which fields do I need".* We included this aspect into REQ2 by splitting annotations into markups ("the word of law") and the attributes ("the word of models").

5 Discussion and Conclusion

This paper contributes a method for the collaborative and semi-automated generation of CMs from LRs in POs. The evaluation revealed that the method can be useful and beneficial for enabling collaboration within and across POs. There is a demand in practice for standardizing and reducing workload as well as redundancy in executing LRs. Yet in all the efforts of standardizing as much as possible, POs also remain autonomous enough to implement their own specifics. Especially smaller POs with less personnel can benefit from such an assistance. Standardized modules also assist in making public services more citizen-centric, as regardless of the process behind the scenes, forms to gather information can be similar [36].

Our method fulfils the requirements for the solution set forth in Sect. 4.1 as explained in the following:

- REQ1 is fulfilled by the introduction of a user-role system. It allows tracking and tracing as well as alleviating the fear of making a wrong choice by having it backchecked by multiple other users.
- REQ2 focusses on markups and annotations to later create models in an automated way. Especially the cascading of linked information was deemed as a major functionality.
- REQ3 as a central piece of the method focusses on creating legally compliant models linked to the LRs. Many tracing needs can be fulfilled by adding attributes like legal source to a model.
- REQ4 is covered by the possibility to export models in frequently used formats.
- REQ5 is fulfilled by the task "Select text(s)" and the subsequent comparison to database entries and highlighting of changes.

Our method and the paper have limitations that lead to suggestions for future work. Up to this stage, we developed and evaluated the concept of our method. A natural next step is the technical realization of the method with a software prototype. Various steps of the method are intended to be supported automatedly, so an implementation would enable a more comprehensive evaluation, both in workshops and test roll-outs in POs. The modular design of the method reflects the stages in which a prototype can be developed and the interfaces required between the modules and to external tools. As of now, the actual modelling step using the text as an input is a black box, however depending on the level of automation and complexity of the text this can be as simple as a form asking for details on the process fragments to fill into a BPMN template, or as complex as a machine learning algorithm combined with natural language processing. Technologies like generative artificial intelligence can be used to retrieve information within a shorter timeframe and a wider reference domain. Even if generative artificial intelligence is not capable of performing a complete automated generation of CM from LR yet, it can be of help by providing a foundation of annotations that human users can build on or by reviewing the annotations made by humans. Additionally, similarity comparisons and recommender systems can point out common or similar definitions, ensuring standardized data to facilitate cross-communication, even if the label on the final appearance is different. For future research sophisticated algorithms can assist in pre-annotating segments based on prior annotations and definitions stored. The evaluations suggested that a sharing system to clone best practices for similar processes could be possible to reduce the workload and redundancy in the operationalisation process for POs.

In conclusion, we propose a method that we hope is helpful in the digitalisation of POs and their services. The need to be compliant to LRs is high in POs and our evaluations indicated that our method has the potential to be a useful artefact in fulfilling this task. It is, however, only one part that can assist in the digitalization of POs. The feedback that we received in the interviews all have one thing in common: Collaboration is the way to go. Pooling resources and knowledge is one future of POs.

Acknowledgements. This project received funding from the Deutsche Forschungsgemeinschaft (DFG, German Research Foundation) – FOR 5393, Project No. 462287308 (SCHO 1965/1-1).

Disclosure of Interests. The authors have no competing interests to declare that are relevant to the content of this article.

References

1. Sevel, M.: Obeying the law. Leg. Theory **24**(3), 191–215 (2018)
2. Dulfer, D., Nijssen, S., Lokin, M.: Developing and maintaining durable specifications for law or regulation based services. In: Ciuciu, I., et al. (eds.) OTM 2015. LNCS, vol. 9416, pp. 169–177. Springer, Heidelberg (2015). https://doi.org/10.1007/978-3-319-26138-6_20
3. Breaux, T.D., Antón, A.I.: Analyzing regulatory rules for privacy and security requirements. IEEE Trans. Software Eng. **34**(1), 5–20 (2008)
4. Callaghan, T., Karch, A., Kroeger, M.: Model state legislation and intergovernmental tensions over the affordable care act, common core, and the second amendment. Publius: J. Fed. **50**(3), 518–539 (2020)
5. van Engers, T., Boer, A., Breuker, J., Valente, A., Winkels, R.: Ontologies in the legal domain. In: Chen, H., et al. (eds.) Digital Government. Integrated Series in Information Systems, vol. 17, pp. 233–261. Springer, Boston (2007). https://doi.org/10.1007/978-0-387-71611-4_13
6. Hoppenbrouwers, S.J.B.A., Proper, H.A., van der Weide, T.P.: A fundamental view on the process of conceptual modeling. In: Hutchison, D., et al. (eds.) ER 2005. LNCS, vol. 3716, pp. 128–143. Springer, Heidelberg (2005). https://doi.org/10.1007/11568322_9
7. Soltana, G.: A model-based framework for legal policy simulation and legal compliance checking. In: Kolovos, D., Chechik, M. (eds.) Proceedings of the Doctoral Symposium at the 18th ACM/IEEE International Conference of Model-Driven Engineering Languages and Systems 2015 (MoDELS 2015). CEUR Workshop Proceedings. Ottawa, CAN (2015)
8. Mendling, J., Recker, J., Reijers, H.A., Leopold, H.: An empirical review of the connection between model viewer characteristics and the comprehension of conceptual process models. Inf. Syst. Front. **21**(5), 1111–1135 (2019)
9. Hevner, A., March, S.T., Park, J., Ram, S.: Design science in information systems research. MIS Q. **28**(1), 75–105 (2004)
10. Greca, I.M., Moreira, M.A.: Mental models, conceptual models, and modelling. Int. J. Sci. Educ. **22**(1), 1–11 (2000)
11. Scholta, H., Niemann, M., Delfmann, P., Räckers, M., Becker, J.: Semi-automatic inductive construction of reference process models that represent best practices in public administrations: a method. Inf. Syst. **84**, 63–87 (2019)
12. Knackstedt, R., Heddier, M., Becker, J.: Conceptual modeling in law: an interdisciplinary research agenda. Commun. Assoc. Inf. Syst. **34**(1), 711–736 (2014)
13. Budhiraja, A., Sharma, K.: Decisions prediction techniques using language processing and learning algorithms. In: Mathur, G., Bundele, M., Tripathi, A., Paprzycki, M. (eds.) ICAIAA 2022. Algorithms for Intelligent Systems, pp. 107–121. Springer, Singapore (2023). https://doi.org/10.1007/978-981-19-7041-2_9
14. Bernstein, A.: Legal corpus linguistics and the half-empirical attitude. Cornell Law Rev. **106**(6), 1397 (2020)
15. Loutsaris, M.A., Charalabidis, Y.: Legal informatics from the aspect of interoperability. In: Charalabidis, Y., Cunha, M.A., Sarantis, D. (eds.) Proceedings of the 13th International Conference on Theory and Practice of Electronic Governance, pp. 731–737. ACM, New York (2020)

16. Schuppan, T., Köhl, S., Off, T.: Vollzugsorientierte Gesetzgebung durch eine Vollzugssimulationsmaschine. NEGZ e.V., Berlin (2018)
17. Pagallo, U.: Network theory and legal information "for" reality: a triple support for deliberation, decision making, and legal expertise. In: Boulet, R., Lajaunie, C., Mazzega, P. (eds.) Law, Public Policies and Complex Systems: Networks in Action. Law, Governance and Technology Series, vol. 42, pp. 267–280. Springer International Publishing, Cham (2019). https://doi.org/10.1007/978-3-030-11506-7_13
18. McLachlan, S., Webley, L.C.: Visualisation of law and legal process: an opportunity missed. Inf. Vis. **20**(2–3), 192–204 (2021)
19. Amaludin, B., Wardika, F.R., Mudana Putra, P.J., Paramartha, I.G.Y.: Analyze the usage of legal definitions in Indonesian regulation using text mining case study: treasury and budget law. In: Schweighofer, E. (ed.) Legal Knowledge and Information Systems. JURIX 2021: The 34th International Conference on Legal Knowledge and Information Systems. Frontiers in Artificial intelligence and Applications, pp. 107–112. IOS Press, Amsterdam (2021)
20. van Engers, T.M., Nijssen, S.: From legislation towards the provision of services. An approach to agile implementation of legislation. In: Kő, A., Francesconi, E. (eds.) EGOVIS 2014. LNCS, vol. 8650, pp. 163–172. Springer, Heidelberg (2014). https://doi.org/10.1007/978-3-319-10178-1_13
21. Breaux, T.D.: Legal Requirements Acquisition for the Specification of Legally Compliant Information Systems. Raleigh, NC, USA (2009)
22. Ciaghi, A., Villafiorita, A., Mattioli, A.: VLPM: a tool to support bpr in public administration. In: Proceedings of the Third International Conference on Digital Society, pp. 289–293. IEEE, Cancun (2009)
23. de Maat, E., Winkels, R., van Engers, T.M.: Making sense of legal texts. In: Grewendorf, G., Rathert, M. (eds.) Formal Linguistics and Law. TiLSM, pp. 225–256. Mouton deGruyter, Berlin (2009)
24. Yip, F.: Semantically enabled applications - a case study in regulatory compliance. Kensington (2012)
25. Wang, X., Vitvar, T., Peristeras, V., Mocan, A., Goudos, S.K., Tarabanis, K.: WSMO-PA: formal specification of public administration service model on semantic web service ontology. In: Proceedings of the HICSS 2007: 40th Annual Hawaii International Conference on System Sciences, p. 96. IEEE, Waikoloa (2007)
26. Morgenstern, L.: Toward automated international law compliance monitoring (TAILCM). Rome, NY, USA (2014)
27. Branting, L.K.: Data-centric and logic-based models for automated legal problem solving. Artif. Intell. Law **25**(1), 5–27 (2017)
28. Casanovas, P., Palmirani, M., Peroni, S., van Engers, T.M., Vitali, F.: Semantic web for the legal domain: the next step. Semant. Web **7**(3), 213–227 (2016)
29. van Engers, T.M., Gerrits, R., Boekenoogen, M.R., Glassée, E., Kordelaar, P.: POWER: using UML/OCL for modeling legislation - an application report. In: Loui, R.P. (ed.) Proceedings of the 8th International Conference on Artificial Intelligence and Law, pp. 157–167. ACM, New York (2001)
30. Nguyen, B.A.P., Scholta, H.: From text to model to execution: a literature review on methods for creating conceptual models from legal regulations. In: ECIS 2024 Proceedings. Association for Information Systems, Paphos (2024)
31. Cherouana, A., Mahdaoui, L., Bellatreche, L., Medjahed, B.: A semantic approach for generating government processes. Int. J. Web Grid Serv. **15**(1), 59–92 (2019)
32. vom Brocke, J., Hevner, A., Maedche, A.: Introduction to Design Science Research. In: vom Brocke, J., Hevner, A., Maedche, A. (eds.) Design Science Research. Cases. Progress in IS, pp. 1–13. Springer, Cham (2020). https://doi.org/10.1007/978-3-030-46781-4_1

33. Peffers, K., Rothenberger, M., Tuunanen, T., Vaezi, R.: Design science research evaluation. In: Peffers, K., Rothenberger, M.A., Kuechler, B. (eds.) DESRIST 2012. LNCS, vol. 7286, pp. 398–410. Springer, Heidelberg (2012). https://doi.org/10.1007/978-3-642-29863-9_29
34. Busch, P.A.: Crafting or mass-producing decisions: technology as professional or managerial imperative in public policy implementation. Inf. Polity **25**(1), 111–128 (2020)
35. Dijkman, R., Dumas, M., van Dongen, B., Käärik, R., Mendling, J.: Similarity of business process models: metrics and evaluation. Inf. Syst. **36**(2), 498–516 (2011)
36. Scholta, H., Balta, D., Räckers, M., Becker, J., Krcmar, H.: Standardization of forms in governments. A meta-model for a reference form modeling language. Bus. Inf. Syst. Eng. **62**(6), 535–560 (2020)

Transparency in Open Government Data Portals: An Assessment of Web Tracking Practices Across Europe

Stefan Stepanovic(✉), Leonardo Mori, Alizée Francey, and Tobias Mettler

Swiss Graduate School of Public Administration, University of Lausanne, Chavannes-près-Renens, Switzerland
stefan.stepanovic@unil.ch

Abstract. Online web analytics and web tracking, including the use of first-party and third-party cookies, are often perceived as a "black box". Both rely on the collection of large amounts of data for various purposes - functional, analytical, and marketing - often without the user's knowledge, for legitimate purposes such as improving the user experience, as well as more controversial reasons such as targeted advertising. This issue is reinforced by Google's dominant position in web analytics, particularly through the widespread integration of Google Analytics (GA) into first-party cookies. At the same time, Europe is witnessing a rise in open government initiatives, particularly in line with the General Data Protection Regulation (GDPR), which aim to increase data transparency and accessibility for individuals. These initiatives often use open government data (OGD) portals as a means to disseminate government information. Our study, therefore, examines such platforms across Europe to determine the prevalence of web tracking activity and Google's potential involvement. Our findings reveal a nuanced use of cookies within OGD portals, characterized by a significant presence of GA cookies. This situation raises debates about privacy (especially in relation to the presence of third-party cookies), transparency, and the possibility of transitioning to more ethically responsible analytics technologies in government digital services. We propose several practical recommendations for governments to improve their privacy efforts, including removing tracking practices, adopting open source analytics solutions, conducting regular audits, and improving public awareness of web tracking practices.

Keywords: Web Analytics · Web Tracking · Privacy · Open Government Data · Google Analytics

1 Introduction

The gradual phase-out of third-party cookies in Google Chrome, starting in 2024, marks a shift in how web analytics operate, driven by the growing focus on online privacy [1]. However, the potential impact on the public sector's expanding online presence remains unclear. This online presence is notably crucial for "open government initiatives" that leverage digital tools to enhance citizen engagement [2, 3]. These initiatives, which include services such as e-citizen portals, offer innovative ways for governments to interact with the population [4].

Historically, web analytics have heavily relied on web tracking, which has been a cornerstone of the multi-billion dollar online advertising industry since the 1990s [5]. Giants like Google and Meta, which captured 56% of global digital ad revenue in 2019, have been primary beneficiaries of this model [5]. This mechanism particularly hinges on web cookies, with two key categories: first-party cookies issued by the website itself, and third-party cookies placed by external entities for cross-site tracking and data sharing [6]. In all cases, such practices support various web functionalities, including personalized content, site analytics, and social media integration [6–8].

In the European Union, the ePrivacy Directive passed in 2002 has been the first regulatory effort to inaugurate the notion of cookie consent [9] and the introduction of the General Data Protection Regulation (GDPR) in 2018 was a significant step towards enforcing strict regulations on data collection and commercial use. This regulation significantly improved the protection of users' privacy against widespread tracking by third parties, in particular through the introduction of consent requirements. However, it is important to acknowledge that third-party cookies are still used on government websites [10]. In 2022, a study found that more than half of government websites in ten G20 countries had third-party cookies in place [11]. Even in Germany, where data protection laws are known to be strict, over 25% of government portals engaged in this practice [11].

In addition, the post-GDPR landscape reveals another paradox: while the regulation aimed to decentralize market power, it appears to have strengthened Google's dominance in various web technology segments [12]. This suggests that GDPR may inadvertently shape the market in favor of larger companies that can leverage first-party cookies data to get around eventual restrictions (such as Google, with Google Analytics) and create new standards (i.e. Google's "Privacy Sandbox", that pushes for the end of third-party cookies) [12, 13]. This development therefore creates a new transition toward more opaque tracking practices, raising concerns about transparency and data stewardship online.

In light of these changes, it is important to examine open government initiatives, which are generally not expected to engage in tracking. Bounded by the GDPR, they are supposed to be guardians of good privacy practices. Additionally, they are often presented as projects that prioritize transparency, collaboration, and public participation in governance. This paper specifically concentrates on Open Government Data (OGD) portals. These portals are indeed a prominent feature of such initiatives: their core objective is to promote transparency by making government data readily available to the public [14].

The central question of our research is: "Do OGD portals track their users?" By investigating the presence of trackers, we seek to provide insight into privacy, transparency, and responsible use of tracking technologies in government services. This is especially important given the growing dominance of Google Analytics in the market.

Our goal is to fully comprehend the current situation, enabling the development of policies that prioritize user interests while respecting privacy concerns. To achieve this, we utilize web scraping techniques to analyze privacy-related data from 35 countries, guided by the 2023 Open Data Maturity Scores. This group includes 27 member states of the European Union (EU), three nations of the European Free Trade Association (EFTA) (Iceland, Norway, and Switzerland), and five candidate countries for EU membership (Bosnia and Herzegovina, Montenegro, Albania, Serbia, and Ukraine).

Our study is structured as follows: first, we provide an overview of the background of the study. We then outline the research methodology and present the findings. Finally, based on these findings, we formulate recommendations and present the limitations of our work.

2 Background

2.1 About First- and Third-Party Cookies

Individuals navigating the online sphere are subject to real-time monitoring across various websites [15, 16], potentially with varying degrees of intensity [17, 18]. In fact, most web services collect significant amounts of personal information from users' online activities through cookies, which constitute the core business of many online companies [6]. Cookies are small text files used to identify a computer using a network. The data stored in the cookies is generated when the computer accesses the network, and it is often labeled with a unique identifier [19]. When the computer emits a request to the website, the cookies can be sent back to the server, which allows the website to recognize the computer thanks to the identifier [19]. This web phenomenon is fueled by diverse enabling techniques [7, 8]; it is ubiquitous on the internet [20] and spans across different websites and devices [21]. Beyond its role in website development and personalization [18, 22], as well as advanced website analytics and integration with social networks [8, 20, 23], it is also commonly used for targeted advertising [8, 24]. However, this information can also be used for other purposes, including price discrimination, personalization of search results, evaluating financial credibility, deciding insurance coverage, surveillance, background scanning, or identity theft [6, 25–27].

Previous studies have highlighted the widespread occurrence of third-party tracking on the internet, which is a form of tracking performed by resources from other services that the one explicitly visited by the users [23]. Hence, a first-party website authorizes a third-party website to gather users' information. Accordingly, third-party tracking enables monitoring of users' activities by services other than the one explicitly visited by the users [6, 20, 23, 28]. This type of tracking poses a significant privacy threat, as it can collect and accumulate vast amounts of personal information (and browsing statistics) from users through many different websites [6, 29]. Moreover, 80% of third-party cookies last more than a month and approximately half of those cookies remain valid for more than a year [27].

Another monitoring practice relies on first-party cookies [29]. While first-party cookies differ from third-party cookies because they aren't automatically transmitted to third parties, it is important to note that, in certain cases, tracking remains feasible [29]. This is due to the fact that any embedded third-party code that operates within the first-party website, may enable to establish or retrieve existing first-party cookies entirely and potentially disclose them to the same or different third parties [29]. While the tracking practices differ, the intrusion into user privacy remains; as tracking performed by a first party is used by a third party to circumvent standard tracking-preventing techniques [30]. However, the invasion of user privacy is even stronger in these cases (i.e., involving tracking through first-party cookies). In fact, in contrast to third-party cookies, which can be easily blocked or deleted without affecting the usability and functionality of the website, first-party cookies are not so easy to block or delete [29]. Moreover, and as noted above, while countermeasures have been implemented to address third-party cookies (e.g., browser blocking third-party cookies), these countermeasures do not target first-party cookies. This is due to the narrative that such monitoring is primarily used to enhance the user experience and improve the site through analytics [30].

More specifically, regarding analytical elements, some of the most popular cookies are Google Analytics (GA) cookies, which aim to analyze website traffic and users' behavior, including all aspects of users' journey through the website (e.g., operating systems, browser versions, session lengths, or page views) [31, 32]. GA is a free web analytical tool hosted by Google that employs first-party cookies to generate detailed statistics based on users' tracking [33, 34]. While GA uses a first-party cookie, set by the website's domain and visible only within that context, it also collects the users' IP as part of its operations, which may constitute individually identifiable information in specific contexts [33]. Accordingly, this raises concerns about data privacy with GA, especially given that all data resides on Google's servers [33], making GA controversial [32, 35].

2.2 About the Rationale Behind OGD Portals

Over the past decades, the advocacy for increased transparency has led to the advent of policy discussions, giving rise to open government initiatives [36–38]. This resulted in a democratized and technological way to enhance the accessibility of government data through OGD portals [39]. OGD portals have thus emerged as flagship open initiatives geared towards achieving a key objective: fostering transparency by disseminating government data [14]. This, in turn, aims to facilitate the accountability of governments and enable the reuse of initially disclosed data, offering social and economic value [14]. While the objective of transparency in OGD is to address information disparity, allowing citizens to observe government activities, transparency also seeks to provide a more accurate account of governments' actions [38]. The discourse on transparency has evolved beyond government actions to encompass organizations that handle diverse data types. This shift is driven by the recognition that transparency not only fosters trust but also seeks to safeguard individuals' right to privacy [40, 41]. This connection between transparency and the right to privacy is the core of regulations like the GDPR, establishing standards for the transparent processing of personal data [10, 42].

Considering, on the one hand, the third-party cookies and, on the other hand, the first-party cookies and the prevalence of GA in analyzing user behavior on websites, it is particularly relevant to investigate these types of trackers on OGD portals. As these portals aim to provide transparent and accessible information, understanding the types of cookies employed, including GA cookies, becomes crucial in assessing the potential impact on websites' transparency and user privacy.

3 Data and Methods

The data we collected are the lists of cookies used by the national OGD portals of the 35 countries included in the 2023 EU open data maturity assessment [43]. In order to obtain them, we have used the Python programming language along with Selenium, an open source project providing several tools for browser automation [44]. Selenium is a well-known tool for such web scraping, widely used in Academia (see, for example, [45] and [46]). We wrote a script that automatically (i) opens the OGD portal homepage, (ii) accepts all cookies (if the consent is requested), and (iii) after waiting for 30 s, collects the name and domain of all the first-party and third-party cookies used by the webpage, as well as the domains of the "frame groups"[1] in which the cookies are sorted. These cookies were collected on March 12th 2024[2].

Once this information was retrieved, we used the cookies' domains to distinguish between first-party and third-party cookies (third-party cookies typically have a different domain associated with them). We then used the Open Cookie Database [47] - an open source initiative that categorize and describe the most commonly used cookies on the Internet -, to gather information about the function of the cookies and the controller of the data they extract. It is worth noting that all the cookies we retrieved from the OGD portals were listed in the Open Cookie Database.

4 Findings

Figure 1 reports the countries where the national OGD portal uses (i) third-party cookies, (ii) GA, (iii) or another web analytics solution. It also indicates (iv) where web cookies were not found during our test and, and finally (v) countries that are non-applicable. The non-applicable countries include Montenegro, whose data portal suffered a cyberattack (as indicated in the 2023 OGD maturity report) and Bosnia and Herzegovina, which is in the process of launching a national OGD portal.

We identified 12 countries that use GA (examples of recurring cookies are _ga, _gid, _gat). Similarly, for web analytics, we categorized 14 countries using alternative platforms: Matomo (example of recurring cookie: _pk; specifically in Austria, France, Germany, Italy, Luxembourg, Netherlands, Poland, Serbia, Sweden, and Switzerland), SiteImprove (example of recurring cookie: nmstat; in Norway and Denmark), and Drupal (example of recurring cookie: has_js; in Belgium and Cyprus). These platforms claim to offer fully GDPR-compliant solutions and position themselves as alternatives to GA. Nonetheless, Drupal (which is a content management system) also presents the option of incorporating a GA module.

---

[1] For an explanation about cookies and frame groups please refer to Google Chrome devtools documentation (reference in bibliography).

[2] The analysis of cookies on OGD portals provides a temporal snapshot that could differ due to site updates, geographical regulations, user behavior, and the browser or device utilized (e.g. system customizations or web tracking countermeasures).

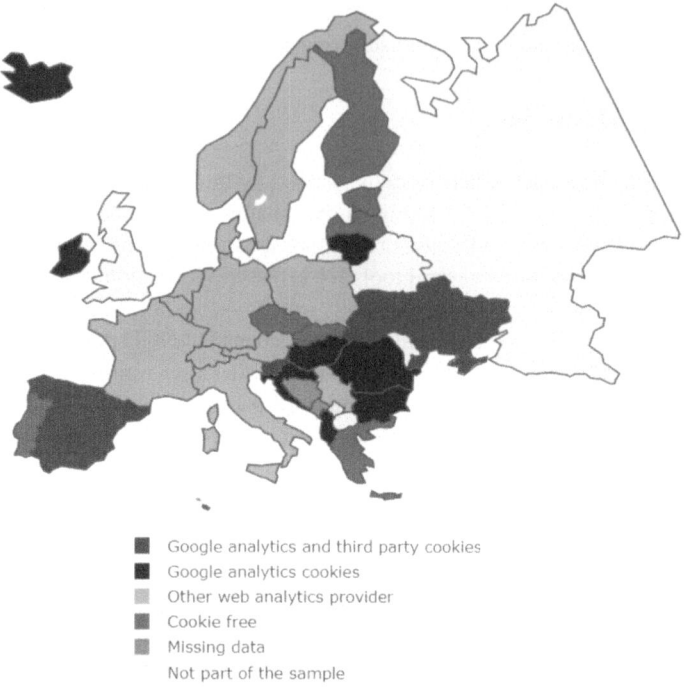

Fig. 1. Map cookies usage on European OGD national platforms

As explained above, companies like Google and Microsoft (with cookies such as _clck, _clsk), offer products allowing their customers to deploy certain web functionalities, analytics or content management systems for their websites. On the other hand, platforms like Matomo and Drupal are open source, meaning that their source code is openly available and not tied to a single vendor. This openness increases transparency and auditability, while providing the ability to ensure that data remains private and secure.

In addition, three countries utilize third-party cookies, that is, domains divergent from the original OGD portal domain displayed on the site. And, ultimately, we have a relatively small proportion of OGD portals that do not have cookies (excluding the countries with missing data, seven out of 33).

Cookies can have different functionalities and are generally grouped in four categories depending on their purpose [47, 48]. The first are functional cookies, which as the name suggests are necessary for the website to work, enabling users to navigate the site and access secure areas. The second are preference cookies, used to remember user choices, hence allowing to personalize user experience. The third are analytics cookies, which collect user behavior aggregated data in order to measure the site performance and thus enable improvements. Last come marketing cookies, that are used to deliver users relevant advertisement, based on their interests and behavior. As it will be discussed later, the distinction between analytics and marketing cookies can sometimes be

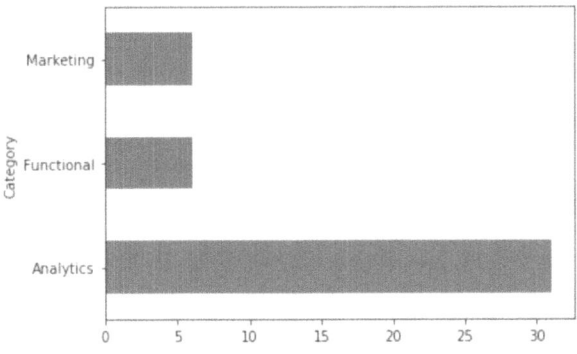

Fig. 2. Cookies categories

blurry[3]. As related by Fig. 2[4], the majority of the cookies we collected fall under the analytics category, primarily related to GA. This is followed by the functional category, and lastly the marketing category (typically third-party cookies). It is important to note that no preference cookies were found.

5 Discussion

Our work aims to shed light on the black box that is web analytics, with a specific emphasis on web tracking, which may be concerning. We particularly focus on the extent of tracking activity within OGD platforms. Although OGD platforms symbolize data transparency and are central to open government initiatives, our findings first reveal the presence of third-party cookies in at least three European countries (out of 35), both within and outside the EU. This raises concerns; even if OGD platforms may not intentionally deploy third-party scripts for tracking, analytics or marketing purposes, the very existence of such scripts can be considered as problematical considering the nature of the websites. This issue persists despite the presence of consent banners, which are frequently overlooked by users (i.e. the *cookie fatigue*: the overwhelming and irritating experience of having to continuously agree to cookies on every new website visited [49, 50]). In fact, there is no technical barrier preventing these third-parties, when they come from private companies, from using analytics data for tracking or profiling (and thus marketing) purposes [10].

However, it is important not solely to blame governments: the inclusion of third-party cookies in OGD platforms may not be intentional [51]. Social media links, video embeds, and "free" software modules can inadvertently introduce tracking cookies due

---

[3] For a more detailed explanation of cookies categories, please visit Cookiepedia: https://cookiepedia.co.uk/classify-cookies.

[4] The counts presented in Fig. 2 represent the number of different types of cookies that were observed, not the total number of cookie. Each cookie type, regardless of its frequency across multiple sites, contributes only once to its category's tally. For instance, despite the _ga cookie appearing on numerous pages, it is counted as a single entry in the analytical category.

to the business models employed by their providers [11], Consequently, OGD platforms that rely on these external applications may inadvertently enable user tracking. Nevertheless, it is certain that the emphasis placed on transparency in the context of OGD leads individuals to not expect to encounter web tracking on OGD portals, especially in EU countries. As we have seen, this expectation is grounded in the fact that these same governments are advocating for anti-tracking measures through regulations such as the GDPR, which mandates data protection both by default and by design (Article 25). Furthermore, these platforms are funded with public money, which should not, in principle, be used to support private or commercial activities such as facilitating tracking or collecting data on citizens [10].

Zooming out, this issue largely stems from outsourcing web development for analytics purposes, coupled with the prevalence of GA in the market [52]. As mentioned, GA holds a significant market share, up to 85.8% of websites by some measures, establishing Google as the de facto standard in web analytics [13, 53]. This is compounded by the dominant position of Google Chrome as a web browser, highlighting the delicate balance between utility and privacy concerns when choosing a web solution [13]. In our analysis, 11 countries (out of 35) employ GA, as revealed by the presence of well-known cookies such as _ga, _gat, and _gid. Yet, our intent is not to criticize these nations; rather, we highlight a grey area where their actions are not necessarily wrongful. In fact, we argue that in the larger scheme of things, giving a single corporation significant influence over data that is primarily intended to benefit citizens can be concerning. Google owns the data collected through its service, which allows it to store, use, and potentially share it [5, 12, 54]. This situation is therefore at odds with the data sovereignty required by European legislation. In 2019, the French data protection authority (CNIL) fined Google 50 million euros for failing to obtain valid consent [54]. Similarly, in 2022, the Austrian data protection authority found Google GA to be in violation of the GDPR [55]. These rulings also stem from the fact that websites using GA are transferring data to the United States, which, along with the potential for data to be reused without consent, is the main issue regarding data protection standards under the GDPR. Nevertheless, we have also seen that GDPR-compliant alternatives are available. This can be in the form of other private companies, such as SiteImprove (storage in the EU, does not include transfer of IP addresses to third countries). Or, it can also be, in a philosophy of transparency, open source alternatives such as Matomo [52]. These opensource alternatives offer fundamentally similar services, including data access and EU data storage, and have been adopted by some countries in our study, such as France and Italy.

Ultimately, only a few countries do not use cookies. It is not a question of better or worse: first-party cookies allow websites to remember individuals' preferences, such as the username and language, for a particular visit. This convenience eliminates the need to re-enter this information while navigating the site and can be used to collect statistics to enhance website features and user experience.

However, it is important to consider whether an OGD portal truly needs to remember all preferences. There exists a delicate balance between what is essential for improving the user experience and what may be considered superfluous.

In sum, while Google's new policy to end third-party cookies aims to enhance user privacy by reducing cross-website tracking, it raises both privacy and antitrust concerns

[12]. It may still permit Google to track users through GA, potentially distorting market competition and favoring Google, leading to increased market concentration [12, 13]. This situation illustrates the complex relationship between competition law and user privacy, and highlights the potential need for regulatory adjustments [13].

6 Implications

6.1 Practical Implications

EU governments are primarily responsible for implementing anti-tracking measures through the GDPR, necessitating data controllers to be accountable [56]. This includes promoting open government initiatives and integrating data protection into all personal data processing activities. We argue that developers of OGD platforms and other government websites may therefore focus on the following areas:

- *Minimize unnecessary tracking:* The necessity of profiling analytics on OGD platforms is questionable. Public sector organizations, unlike private companies, have a distinct mission that does not prioritize maximizing website traffic. Instead, their primary objective should be enhancing citizen experience by ensuring easy access to required services [52]. If some analytics are deemed necessary to understand user behavior and identify improvements, a privacy-friendly, citizen-centric approach should be adopted for data collection.
- *Adopt open source products:* In line with the above point, to mitigate risks, OGD platform managers may adopt open standards and open source software. This approach reduces reliance on proprietary services and technologies and is consistent with the transparency ethos of open government initiatives [52]. Finally, a relatively simple yet radical solution is to remove (when in doubt) the highly analytical components of the code altogether [57].
- *Conduct regular audits:* Regulators in each country need to regularly conduct detailed audits to monitor third-party tracking on government websites [11, 58], with the goal of quickly removing trackers. As detailed in this study, it is relatively easy for government agencies, civil society, and independent researchers to help conduct audits [58]. They can create inventories and report tracking incidents. Tools such as the Open Cookie Database [47] can also help with isolated and online checks.
- *Increase awareness about web tracking*: Data protection authorities should raise awareness about the risks of web tracking and the general loss of control over data use [59]. In fact, privacy policies on websites are often vague, broad, and lengthy, ultimately leading to the risk of uninformed online decisions [11, 13, 58]. For example, best practices for consent banners (although not the focus of this study) include: providing clear choices, securing consent for non-essential cookies, communicating data use transparently, documenting consent, maintaining accessibility without certain cookies, and making it easy to withdraw consent [49, 50]. This means avoiding pre-checked boxes, hidden opt-out options, multiple clicks, and tracking without or prior to consent [50].

Ultimately, individuals can also protect their privacy by using browser extensions like Privacy Badger or uBlock to increase control over their personal information [60] and contribute to broader awareness and action against intrusive web tracking.

6.2 Theoretical Implications

The theoretical implications of the study provide insights into the intersection of government transparency, web tracking, and privacy concerns. We reveal the controllers of web tracker mechanisms on OGD platforms and suggest that transparency could be enhanced by using open-source analytics software (when necessary). This also highlights the tension between privacy and utility in practice [13, 53]. Organizations must balance operational efficiency and innovation with ethical considerations of privacy and data protection. This tension is further amplified by high-level regulations such as GDPR. Although the objectives and rationale of the law are clear, there are minimal procedures at a lower level (i.e. operational level) on how to implement it, as illustrated in our present study within the context of privacy and web tracking [61, 62]. However, this situation also reflects the relative novelty of the regulation, implying that practices will evolve, and increasingly evidence-based methodologies will emerge for effective compliance.

Finally, the methodological approach adopted in this study - combining web scraping techniques with an analysis of cookie functionalities/data controllers - provides a replicable design for future research in this field. This approach can be adapted and extended to other contexts, allowing scholars to conduct audits of web tracking practices across different sectors and regions.

7 Limitations and Concluding Remarks

Our study has primarily examined cookies as a component of online tracking. However, beyond the scope of first-party and third-party cookies, a variety of techniques are employed to monitor and analyze online user behavior. These techniques leverage different features of web technologies for tracking purposes. Key methods include, for instance, web beacons and fingerprinting [51, 63]. Web beacons are invisible images or small code snippets embedded in websites and emails, enabling the tracking of user activities (i.e. website visits). This is achieved by making a request to a server for the image, which logs this request [63]. Fingerprinting collects data on a user's device and browser configurations [64], such as screen resolution, operating system, and font types, to generate a unique identifier for the device [65]. Fingerprinting can track users across different browsing sessions, even in private mode, as it does not depend on cookies [64]. Thus, our findings likely underrepresent the full extent of web tracking's pervasiveness, which concretely highlights a need for further investigation.

An additional limitation of our research is that we did not focus on analyzing cookie consent mechanisms, which are closely related to GDPR requirements for privacy management and online personal data protection. This includes the variability and complexity that users encounter when managing their preferences, especially when opting out (which is a major concern under the GDPR [51]). Such variability in user interface design for cookie management may indeed alter the opt-out process by making it complex and discouraging [50, 51]. And, as we have outlined, this can lead to "cookie fatigue", i.e., the annoyance experienced by users due to the frequent requests for cookie permissions on websites [66].

Due to the format of the study, we also did not perform an in-depth content analysis of the cookies. Examining the content of these cookies could reveal whether they store information that could potentially be used for tracking, such as unique identifiers. We recommend that future research include this analysis to provide a more detailed understanding of the tracking capabilities of specific cookies.

In conclusion, our findings reveal a relatively widespread use of ready-made analytical solutions, first among all GA. Even though such products certainly provide good quality services and have somehow become a standard practice in web development, we believe that it is important for governments to consider the consequences of using these analytical suites. Indeed, using such products can allow foreign big tech companies to gain access to European citizens online behavior data, and to stock it outside Europe, which does not align with the GDPR [54, 55]. Additionally, besides legal considerations, it is often impossible to clearly rule out the possibility that data collected with analytical cookies will not be used for marketing purposes [13, 35]. Therefore, it is essential for governments to align their practices in developing OGD portals with the privacy recommendations they issue. This ensures coherence and transparency, which are the leading motivations behind OGD initiatives [14]. Although digitalization is often viewed as a means of improving the efficiency of public administration, it should not lose sight of its primary goal of enhancing the quality of public services and making them accessible and fair to all citizens.

Acknowledgments. This study was funded by the Swiss National Science Foundation (grant number 212637). We would like to give a special thanks to Hugo Hueber, research engineer at the University of Lausanne, for his support.

Disclosure of Interests. The authors have no competing interests to declare that are relevant to the content of this article.

References

1. Cooper, D.A., Yalcin, T., Nistor, C., Macrini, M., Pehlivan, E.: Privacy considerations for online advertising: a stakeholder's perspective to programmatic advertising. J. Consum. Mark. **40**(2), 235–247 (2023)
2. Soguel, N., Bundi, P., Mettler, T., Weerts, S.: Comprendre et concevoir l'administration publique, 1st edn., EPFL Press (2023)
3. Wirtz, B.W., Weyerer, J.C., Becker, M., Müller, W.M.: Open government data: a systematic literature review of empirical research. Electron. Mark. **32**(4), 2381–2404 (2022)
4. Schedler, K., Guenduez, A.A., Frischknecht, R.: How smart can government be? Exploring barriers to the adoption of smart government. Inf. Polity **24**(1), 3–20 (2019)
5. Johnson, G.: Economic research on privacy regulation: lessons from the GDPR and beyond (2022)
6. Bujlow, T., Carela-Español, V., Solé-Pareta, J., Barlet-Ros, P.: A survey on web tracking: mechanisms, implications, and defenses. Proc. IEEE **105**(8), 1–34 (2017)
7. Besson, F., Bielova, N., Jensen, T.: Hybrid information flow monitoring against web tracking. In: 26th Computer Security Foundations Symposium, pp. 240–254. IEEE, New Orleans (2014)
8. Sanchez-Rola, I., Ugarte-Pedrerp, X., Santos, I., Bringas, P.G.: The web is watching you: a comprehensive review of web-tracking techniques and countermeasures. Log. J. IGPL **25**(1), 18–29 (2016)

9. Debusseré, F.: The EU e-privacy directive: a monstrous attempt to starve the cookie monster? Int. J. Law Inf. Technol. **13**(1), 70–97 (2005)
10. Samarasinghe, N., Adhikari, A., Mannan, M., Youssef, A.: Et tu, brute? Privacy analysis of government websites and mobile apps. In: ACM Web Conference, pp. 564–575. ACM, Lyon, France (2022)
11. Gotze, M., Matic, S., Iordanou, C., Smaragdakis, G., Laoutaris, N.: Measuring web cookies in governmental websites. In: Proceedings of the 14th ACM Web Science Conference 2022, Barcelona, Spain, pp. 44–54 (2022)
12. Peukert, C., Bechtold, S., Batikas, M., Kretschmer, T.: Regulatory spillovers and data governance: evidence from the GDPR. Mark. Sci. **41**(4), 746–768 (2022)
13. Geradin, D., Katsifis, D., Karanikioti, T.: Google as a de facto privacy regulator: analyzing Chrome's removal of third-party cookies from an antitrust perspective (2020)
14. Lourenço, R.P.: An analysis of open government portals: a perspective of transparency for accountability. Gov. Inf. Q. **32**(3), 323–332 (2015)
15. Gomer, R., Rodrigues, E.M., Milic-Frayling, N., Schraefel, M.C.: Network analysis of third party tracking: user exposure to tracking cookies through search. In: IEEE/WIC/ACM (ed.) International Joint Conferences on Web Intelligence (WI) and Intelligent Agent Technologies (IAT), vol. 1, pp. 549–556. IEEE, Atlanta (2013)
16. Falahrastegar, M., Haddadi, H., Uhlig, S., Mortier, R.: Tracking personal identifiers across the web. In: In: Karagiannis, T., Dimitropoulos, X. (eds.) PMS 2016. LNCS, vol. 9631 pp. 30–41. Springer, Cham (2016). https://doi.org/10.1007/978-3-319-30505-9_3
17. Ermakova, T., Hohensee, A., Orlamünde, I., Fabian, B.: Privacy-invading mechanisms in e-commerce - a case study on German tourism websites. Int. J. Netw. Virtual Organ. **20**(2), 105–126 (2017)
18. Ermakova, T., Fabian, B., Bender, B., Klimek, K.: Web tracking – a literature review on the state of research. In: 51st Hawaii International Conference on System Sciences, Hilton Waikoloa Village, Hawaii, pp. 4732–4741 (2018)
19. What are Cookies?. https://www.kaspersky.com/resource-center/definitions/cookies. Accessed May 2024
20. Roesner, F., Kohna, T., Wetherall, D.: Detecting and defending against third-party tracking on the web. In: 10th International Conference on Web and Social Media, San Jose, USA, pp. 155–168 (2012)
21. Brookman, J., Rouge, P., Alva, A., Yeung, C.: Cross-device tracking: measurement and disclosure, privacy enhance technologies, pp. 133–148 (2017)
22. Fourie, I., Bothma, T.: Information seeking: an overview of web tracking and the criteria for tracking software. In: Aslib (Ed.), pp. 264–284 (2007)
23. Mayer, J.R., Mitchell, J.C.: Third-party web tracking: policy and technology. In: IEEE Symposium on Security and Privacy, pp. 413–427. IEEE, San Francisco (2012)
24. Parra-Arnau, J.: Pay-per tracking: a collaborative masking model for web browsing. Inf. Sci. **1**(385), 96–124 (2017)
25. Mikians, J., Gyarmati, L., Erramilli, V., Laoutaris, N.: Detecting price and search discrimination on the internet. In: 11th ACM Workshop on Hot Topics in Network, ACM, Washington, USA, pp. 79–84 (2012)
26. Hannak, A., Soeller, G., Lazer, D., Mislove, A., Wilson, C.: Measuring price discrimination and steering on E-commerce web sites. In: Internet Measurement Confer, pp. 305–318. ACM, Vancouver (2014)
27. Samarasinghe, N., Mannan, M.: Towards a global perspective on web tracking. Comput. Secur. **87**(101569), 1–13 (2019)
28. Li, T.-C., Hang, H., Faloutsos, M., Efstathopoulos, P.: TrackAdvisor: taking back browsing privacy from third-party trackers. In: Mirkovic, J., Liu, Y. (eds.) PAM 2015. LNCS, vol. 8995, pp. 277–289. Springer, Cham (2015). https://doi.org/10.1007/978-3-319-15509-8_21

29. Chen, Q., Ilia, P., Polychronakis, M., Kapravelos, A.: Cookie swap party: abusing first-party cookies for web tracking. In: 2021 ACM Web Conference, Ljubljana, Slovenia, pp. 2117–2129 (2021)
30. Demir, N., Theis, D., Urban, Z., Pohlmann, N.: Towards understanding first-party cookie tracking in the field. arXiv preprint arXiv:2202.01498, pp. 1–20 (2022)
31. Pantelic, O., Jovic, K., Krstovic, S.: Cookies implementation analysis and the impact on user privacy regarding GDPR and CCPA regulations. Sustainability **14**(9), 1–14 (2022)
32. The End of Google Analytics in Europe?. https://www.activemind.legal/guides/google-analytics/. Accessed 12 Jan 2024
33. Loftus, W.: Demonstrating success: web analytics and continuous improvement. J. Web Librariansh. **6**(1), 45–55 (2012)
34. Plaza, B.: Monitoring Web Traffic Source Effectiveness with Google Analytics: An Experiment with Time Series, pp. 474–482. Aslib, Emerald Group Publishing Limited (2009)
35. Is Google Analytics 4 GDPR-compliant?. https://usercentrics.com/knowledge-hub/google-analytics-and-gdpr-compliance-rulings. Accessed 12 Jan 2024
36. Bertot, J.C., Jaeger, P.T., Grimes, J.M.: Using ICTs to create a culture of transparency: E-government and social media as openess and anti-corruption tools for societies. Gov. Inf. Q. **27**(3), 264–271 (2010)
37. McDermott, P.: Building open government. Gov. Inf. Q. **27**(4), 401–413 (2010)
38. Matheus, R., Janssen, M.: A systematic literature study to unravel transparency enabled by open government data: the window theory. Perform. Manag. Rev. **43**(3), 503–534 (2020)
39. Open Government Data - What is Open Government Data?. https://www.oecd.org/gov/digital-government/open-government-data.htm. Accessed 14 Mar 2022
40. Nougrères, A.B.: Privacy is key in processing personal data by AI: UN expert, united nations, online (2023)
41. Tolbert, C.J., Mossberger, K.: The effects of E-government on trust and confidence in government. Public Adm. Rev. **66**(3), 354–369 (2006)
42. Official Journal of the European Union, Regulation (EU) 2016/679 of the European Parliament and the Council of 27 April 2016 on the protection of natural persons with regard to the processing of personal data and on the free movement of such data, and repealing Directive 95/46/EC (General Data Protection Regulation), pp. 1–88 (2016)
43. data.europe.eu. 2023 Open Data Maturity Report, p. 152 (2023)
44. Selenium documentation. https://www.selenium.dev/. Accessed 12 Mar 2024
45. Rasaii, A., Singh, S., Gosain, D., Gasser, O: Exploring the cookieverse: a multi-perspective analysis of web cookies. In: Brunstrom, A., Flores, M., Fiore, M. (eds.) PAM 2023. LNCS, vol. 13882, pp. 623–651. Springer, Cham (2023). https://doi.org/10.1007/978-3-031-28486-1_26
46. Englehardt, S., et al.: Cookies that give you away: the surveillance implications of web tracking. In: Proceedings of the 24th International Conference on World Wide Web, Florence, Italy, pp. 289–299 (2015)
47. Open Cookie Database. https://github.com/jkwakman/Open-Cookie-Database?tab=readme-ov-file. Accessed 12 Mar 2024
48. Kretschmer, M., Pennekamp, J., Wehrle, K.: Cookie banners and privacy policies: measuring the impact of the GDPR on the web. ACM Trans. Web (TWEB) **15**(4), 1–42 (2021)
49. Pantelic, O., Jovic, K., Krstovic, S.: Cookies implementation analysis and the impact on user privacy regarding GDPR and CCPA regulations. Sustainability **14**(9), 5015 (2022)
50. Habib, H., Li, M., Young, E., Cranor, L.: "Okay, whatever": an evaluation of cookie consent interfaces. In: Proceedings of the 2022 CHI Conference on Human Factors in Computing Systems, pp. 1–27 (2022)

51. Papadogiannakis, E., Papadopoulos, P., Kourtellis, N., Markatos, E.P.: User tracking in the post-cookie era: How websites bypass GDPR consent to track users. In: 2021 Proceedings of the web Conference, pp. 2130–2141 (2021)
52. Gamalielsson, J., et al.: Towards open government through open source software for web analytics: the case of Matomo. JeDEM-eJ. eDemocr. Open Govern. **13**(2), 133–153 (2021)
53. Alby, T.: Popular, but hardly used: has Google Analytics been to the detriment of web Analytics?. In: 2023 Proceedings of the 15th ACM Web Science Conference, pp. 304–311 (2023)
54. Urban, T., Tatang, D., Degeling, M., Holz, T., Pohlmann, N.: Measuring the impact of the GDPR on data sharing in ad networks. In: Proceedings of the 15th ACM Asia Conference on Computer and Communications Security, pp. 222–235 (2020)
55. Winklbauer, S., Horner, R.: Austrian DPA decides EU-US data transfer through the use of google analytics to be unlawful. Eur. Data Prot. L. Rev. **8**, 78 (2022)
56. Kollnig, K., Shuba, A., Van Kleek, M., Binns, R., Shadbolt, N.: Goodbye tracking? Impact of iOS app tracking transparency and privacy labels. In: Proceedings of the 2022 ACM Conference on Fairness, Accountability, and Transparency, Seoul, South Korea, pp. 508–520 (2022)
57. Tahaei, M., Li, T., Vaniea, K.: Understanding privacy-related advice on stack overflow. Priv. Enhanc. Technol. **2022**(2), 114–131 (2022)
58. Libert, T.: An automated approach to auditing disclosure of third-party data collection in website privacy policies. In: Proceedings of the 2018 World Wide Web Conference, pp. 207–216 (2018)
59. The Federal Council, Federal Data Protection and Information Commisionner. https://www.edoeb.admin.ch/edoeb/en/home/datenschutz/grundlagen.html. Accessed January 2024
60. Borgolte, K., Feamster, N.: Understanding the performance costs and benefits of privacy-focused browser extensions. In: Proceedings of the Web Conference 2020, Taipei, Taiwan, pp. 2275–2286 (2020)
61. Várkonyi, G.G., Gradišek, A.: Data protection impact assessment case study for a research project using artificial intelligence on patient data. Informatica **44**(4), 1–10 (2020)
62. Karami, F., Basin, D., Johnsen, E.B., DPL: a language for GDPR enforcement. In: 2022 IEEE 35th Computer Security Foundations Symposium (CSF), pp. 112–129. IEEE, Haifa (2022)
63. Kashi, E., Zavou, A.: Did I agree to this? Silent tracking through beacons. In: Moallem, A. (eds) HCII 2020. LNCS, vol. 12210, pp. 427–444. Springer, Cham (2020). https://doi.org/10.1007/978-3-030-50309-3_28
64. Al-Fannah, N.M., Mitchell, C.: Too little too late: can we control browser fingerprinting? J. Intellect. Cap. **21**(2), 165–180 (2020)
65. Acar, G., et al.: The web never forgets: Persistent tracking mechanisms in the wild. In: Proceedings of the 2014 ACM SIGSAC Conference on Computer and Communications Security, New York, USA, pp. 674–689 (2014)
66. Bollinger, D., Kubicek, K., Cotrini, C., Basin, D.: Automating cookie consent and {GDPR} violation detection. In: 31st USENIX Security Symposium (USENIX Security 2022), Boston, USA, pp. 2893–2910 (2022)

Understanding Trust Frameworks: Goals and Components Identified Through a Case Study

Louise van der Peet[1(✉)], Nitesh Bharosa[1], Sander Dijkhuis[2], and Marijn Janssen[1]

[1] Delft University of Technology, Delft, The Netherlands
l.vanderpeet@tudelft.nl
[2] Cleverbase, The Hague, The Netherlands

Abstract. Amidst increasing online data sharing among organisations, there is a growing need for interoperability and trust in the digital space. When there is no infrastructure provider for sharing information (e.g. by Big Tech players and/or government-owned infrastructures), public and private actors must figure out how to reach agreements about the technical specifications and data infrastructure components to facilitate inter-organizational collaboration. This paper zooms in on the empirical phenomenon of trust frameworks emerging in practice. The main research question is twofold: (1) what are the goals actors strive for with trust frameworks and (2) which components are developed for achieving these goals? Drawing on previous literature and a case study approach, interoperability, certainty, efficiency, and security emerge as goals of trust frameworks. As for the second question, we draft an exhaustive diagram of components from both the literature and our case study. This explorative research lays the foundation for future research into trust frameworks as a major change in traditional approaches to cross sector data exchange.

Keywords: Trust frameworks · Digital governance · Data interoperability

1 Introduction

Globally, digital data sharing has reached unprecedented heights, organisations and individuals (unintendedly) share massive amounts of data. There is an ongoing steep data growth going on, with only 2 zettabytes in 2010, and in 2025 it is projected to be 181 zettabytes [23]. As shared data volume and sensitivity continue to grow, so does the need for robust mechanisms to ensure secure, efficient and trustworthy data sharing across various sectors and organisations. This is especially true for situations where sharing incorrect data can have legal consequences. Examples include applying for public services, filing corporate reports and buying a house [7]. In such cases, all actors seek legal certainty, for instance about the identity of the supplier and receiver of data, as well as the confidentiality, integrity and availability of digital infrastructure components. When there is no infrastructure provider (e.g. by Big Tech players and/or government-owned infrastructures), organisations must themselves make to agreements about various specifications and data infrastructure components needed for interoperability and trust for inter-organizational data sharing. In other words, they are developing trust frameworks which guide information sharing among public and private organisations. The need for interoperability across organisations and sectors is recognized on a broader scale, as shown by the interoperability initiative in the European Union, the European Interoperability Act. This legislation aims to promote a more open and secure digital space, encouraging cooperation across borders and sectors [20].

Against this background, this paper zooms in on the empirical phenomenon of trust frameworks. Trust frameworks are capturing how actors seek to build cross-organisational and cross-sectoral interoperability and trust.

An example of such a framework in action is MedMij [13], a Dutch initiative aimed at establishing a secure and reliable ecosystem for health data exchange. MedMij serves as a set of standards and agreements designed to ensure that personal health data can be shared securely between healthcare providers' and patients' personal health environments. Prior to MedMij, the healthcare sector faced significant challenges due to the disparate methods of data sharing among various healthcare providers. These inconsistencies can lead to fragmented patient information, inefficiencies, and increased risks to patient privacy [11]. MedMij addresses these problems by providing a unified framework that standardizes data exchange processes. This framework enables patients to gain control over their health data, merging information from various sources into one single location. Medmij was developed through collaboration between healthcare providers, app developers, patient representatives and personal health environment providers. These efforts have resulted in a secure, interoperable framework that empowers patients to manage their health data while promoting efficiency and privacy.

While organizations are increasingly incorporating trust frameworks, such as the MedMij framework, there is little scientific literature providing a conceptualization of trust frameworks, and in the literature that exists, there is no complete consensus on the term (Sect. 3 provides more details).

This explorative paper aims to introduce a clearer understanding of trust frameworks. Our goal is to comprehensively conceptualize trust frameworks, including components and goals. Accordingly, the main research question is twofold: (1) what are the goals actors strive for with trust frameworks and (2) which components are developed for achieving these goals? By offering a conceptualization informed by literature and empirical evidence, this research aims to contribute to theory building on trust frameworks, advancing academic scholarship and practical applications. This paper proceeds as follows. Section 2 presents the research approach followed and Sect. 3 provides a conceptualization of trust frameworks. Section 4 reveals the findings of the empirical case study. Section 5 concludes this paper and presents directions for future research.

2 Research Approach

The objective of this explorative paper is to gain a better understanding of trust frameworks, particularly the definition, components and goals. We follow a three-step approach in order to achieve the objective. First, we conduct a systematic literature review on the concept of trust frameworks, allowing for the development of a lens for investigating the case study. The main goals of this literature research is to find components of trust frameworks in the literature, and a robust definition for the term. We employed thematic analysis to identify and cluster key components into broader categories, ensuring a systematic and rigorous understanding of trust frameworks. Section 3 provides an overview of the findings of the literature review.

Second, we conduct an empirical case study that zooms in on the ongoing development of a trust framework called Trusted Information Partners (TIP) [25]. TIP is a public-private system of agreements for the exchange of digital data, primarily with significant financial or legal impact. Yin [27] proposes that case studies contribute to the examination of contemporary phenomena within their natural context, especially in new phenomena where existing theory might be limited; and that a single use case study is well-suited for exploratory research aimed at theory development. The TIP case study was chosen due to its widely collaborative nature, involvement in the public sector, and because it is used for the purpose of data sharing, making the case relevant for the research goals. The qualitative data was collected through semi-structured interviews, with representatives from different organisations.

We interviewed a total of eight experts involved in the case study. The interviewees were gathered through the network of the researchers, the criteria for the interviewees being that they are actively involved in the development of TIP; and have expertise on some aspect of digital data sharing. The respondent description can be found in Table 3. Each interview lasted around 60 min. Three main goals were pursued in the interviews: (1) find a common definition for trust frameworks, (2) gather the essential components for trust frameworks, and (3) find the goals that are pursued using trust frameworks. The resulting interviews were recorded, transcribed, validated by the respondents, and analysed by comparing common themes. This analysis involved thematic coding, where each interview transcript was examined line-by-line to identify and categorize mentioned goals and components, allowing for the systematic identification of recurring themes.

Lastly, the findings from the literature review are compared with the findings from the interview. The main findings of this comparison are captured in Fig. 1. This leads to a greater understanding of the components and definition of trust frameworks.

3 Conceptualizing Trust Frameworks

We searched the literature for examples of trust frameworks, in order to establish a definition and conceptualization. A Scopus 'title-keyword-abstract' search for the keywords *'trust framework'* and *'governance'* in November 2023 reveals 37 papers.

Alternative terms such as *'trust model'*, *'trust scheme'*, and *'trust protocol'* were considered but ultimately excluded from the primary search. The reasons for this exclusion are twofold: (1) these terms are often related to quantifying trust for better decision making, rather than creating a practical framework that can be used without trust among the participants; and (2) including these terms leads to more irrelevant results in the systematic review. For instance, the term *'trust model'* refers to "methods on how to model and quantify trust with sufficient detail and context-based adequateness" [5]. 'Trust schemes' help automate trust decisions by leveraging technical standards, legal regulations, and infrastructure, highlighted by More [15]. Lastly, 'trust protocol' refers to a method for modeling indirect trust [21]. Including these terms would thus dilute the focus of our review and lead to less relevant literature being considered.

In the set of papers, only one contains a definition for the term trust framework. Brewer et al. [3] take a definition from the white paper of the National Institute of Standards and Technology (NIST) [24] where trust frameworks are defined as following:

> "a generic term used to describe a legally enforceable set of specifications, rules, and agreements that govern a multi-party system established for a common purpose, designed for conducting specific types of transactions among a community of participants, and bound by a common set of requirements".

All other papers in the review use the term one or multiple times throughout the paper, but never refer to its definition. This confirms our starting point, that there is no universal definition for trust frameworks. Furthermore, the term is also used to describe a framework for conceptualizing trust, rather than the sort of practical framework that builds cross-organisational and cross-sectoral interoperability and trust that we aim to discuss [6,9,14,18]. In this analysis we will mostly focus on the practical perspectives on trust frameworks.

In the literature that describes practical frameworks, we identify several common components of trust frameworks that recur across the literature as outlined in Table 4. In this context, the term 'components' is defined as something needed in order to facilitate information sharing or the governance of information sharing. We extracted all components from the literature, and group

Table 1. Categories of trust framework components in the literature

Study	Technical specifications	Governance specification	Operational requirements	Legal requirements
[1]	✓	✓		
[2]	✓	✓	✓	✓
[3]	✓	✓	✓	✓
[4]	✓			
[8]		✓		
[10]	✓		✓	
[12]	✓		✓	✓
[16]		✓	✓	✓
[19]	✓	✓	✓	✓
[22]	✓	✓		

these into categories: similar components were clustered together based on their purpose and through this grouping patterns emerge, allowing the discerning of categories. Four categories emerge from this analysis: operational requirements, legal requirements, governance, and technical implementation (Table 1).

- **Operational Requirements:** These are the procedural and protocol-driven aspects critical for the daily functionalities and security measures of the trust framework, including data operations and technical specifications.
- **Legal Requirements:** This category represents the compliance and regulatory framework, ensuring alignment with legal standards and practices through e.g. audit schemes and terms of service.
- **Governance:** Governance components outline the organizational structure and the distribution of roles within the framework, detailing the mechanisms for auditing, validation, and collaboration among stakeholders.
- **Technical Implementation:** Most of the literature relies on technical specifications to model a trust framework. This occurs in various forms, from the implementation of user dashboards [19] to the application of blockchain methods [22].

Furthermore, in the literature we found a diverse application of trust frameworks across several sectors. For example improving data exchange in the food and agriculture sector [3, 19], and improving trust among users of virtual environments [4, 12]. The wide range of sectors applying trust frameworks underscores the universal relevance of trust frameworks in fostering trust and security across various industries.

We further identify the goals of the trust frameworks. These can be summarized into five categories:

- **Security and privacy:** nine out of ten of the relevant papers mention security and privacy as a goal of the trust framework. More specifically, privacy of

personal data [3,8,10,12,16], and the security principles of the CIA triad (confidentiality [16], intergrity [1,10], availability [12,16,22]) are common goals in the literature.
- **Certainty and compliance:** Increasing certainty and compliance is another goal. Certainty in the form of accountability [2,16,19], and in the form of reliable information [4,10,22] were most common in the literature. Compliance to privacy regulations [3,16] and legal certainty [2] could be important, with frameworks designed to align with existing laws and standards to facilitate adoption and integration into current systems.
- **Societal impact:** Trust frameworks in different sectors have different goals in terms of societal benefits. From the improvement of food safety and quality [3], to maintaining free movement during the COVID-19 pandemic [8]. More common goals in this category are: users' control over their personal data [9,10,19,22]; establishing ethics and shared values [2,12,22]; and increasing transparency [19,22].
- **Interoperability and scalability:** Interoperability is a goal that emphasizes the need for systems and technologies to work together seamlessly, it is specifically mentioned as a goal in two previous works [8,16]. Scalability [22] and multi-lateral data exchange [3] are found as goals as well.
- **Efficiency:** Efficiency is mentioned only by one study as a goal [3]. This trust framework promotes efficiency by means of supply-chain efficiency, and by unlocking the full potential of already-existing technologies.

An overview of which relevant work contains which categories of goals can be found in Table 2.

Table 2. Categories of trust frameworks' goals in the literature

	Security & privacy	Certainty & compliance	Societal impact	Interoperability & scalability	Efficiency
[1]	✓				
[2]	✓	✓	✓		
[3]	✓	✓	✓	✓	✓
[4]		✓	✓		
[8]	✓		✓	✓	
[10]	✓	✓	✓		
[12]	✓		✓		
[16]	✓	✓		✓	
[19]	✓	✓	✓		
[22]	✓	✓	✓	✓	

A unified understanding of trust frameworks is still missing, underscoring the need for a more cohesive conceptualization.

4 Case Study: Trusted Information Partners

4.1 Background

Trusted Information Partners (TIP) is an initiative consisting of public and private parties in the Netherlands, where the goal is to increase ease and trustworthiness of online interactions for citizens, businesses and government. The collaborating partners share a vision that once citizens and entrepreneurs have access to a high level of assurance; electronic identities; functions for personal data management; and exchange within relevant legal frameworks, eSociety would function without paperwork. The resulting product of this collaboration is a cross-domain public-private trust framework. For this, components are specified and governed regarding the following topics:

- Identities at a high level of assurance (conform eIDAS)
- Qualified trust services (conform eIDAS)
- Methodology for legal representation with a high level of assurance
- Methodology for funding of collective functionalities and (maintenance of) the trust framework
- Shared digital infrastructure for information exchange
- Trusted registration and publication of service and chain specifications
- Discovery of shared information services
- Payment system for services delivered within the ecosystem.

The trust framework is an implementation of the European eIDAS regulation (EU) 910/2014 [26], establishing trust frameworks for electronic identification and trust services in the internal market. As stated in the eIDAS recitals, there is a need for online trust and legal certainty.

> "(1) Building trust in the online environment is key to economic and societal development. Lack of trust, in particular because of a perceived lack of legal certainty, makes consumers, businesses and public authorities hesitate to carry out transactions electronically and to adopt new services".
> "(2) This Regulation seeks to enhance trust in electronic transactions in the internal market by providing a common foundation for secure electronic interaction between citizens, businesses and public authorities, thereby increasing the effectiveness of public and private online services, electronic business and electronic commerce in the Union".

Whereas eIDAS establishes minimum legal and technical requirements and standards to enable electronic transactions, the regulation does not replace member state laws and agreements. The way electronic identification and trust services are applied differs per member state. The partners in TIP aim to leverage the legal and technical standards of eIDAS in a public-private collaboration to design, govern and contribute to the adoption of qualified information exchange on a shared infrastructure in a common trust framework.

4.2 Interview Results

We have interviewed eight experts who are all currently part of the TIP collaboration. Table 3 provides an overview of interview respondents. These respondents were selected based on their expertise in developing components. Seven out of the eight respondents have been involved in the implementation of at least one other trust framework. In this section we will discuss how respondents define trust frameworks and what they see as the goals for trust frameworks.

Table 3. Overview of interview respondents

Respondent	Role	Expertise
1	Product owner data exchange at public organisation	maintenance and standardisation of trust frameworks, governance
2	Customer journey expert at private organisation	finance, digital innovation
3	Advisor for digital society at public organisation	governance, standardised information exchange development of trust frameworks
4	Policy officer at public organisation	control over data, data sharing, government transparency, data governance, policy frameworks
5	Product owner and security officer at private association	tech, finance, efficiency, security and privacy, enterprise architecture, requirements
6	Business architect at private organisation	compliance, process design, management, data sharing, cyber security
7	CTO at private organisation	taxonomies, technical strategies and operations, standard reporting
8	Project manager innovation at private organisation and ecosystem developer via foundation	ecosystem development, control over data, communication, strategy, standardisation

Definition. When discussing definitions of trust frameworks, respondent 1 described the need for an independent regulatory organisation (possibly the government), respondent 3 mentioned the introduction of processes or technology in a standardised way, and reliable communication, while respondent 6 mentioned multiple parties in a chain working together to introduce agreements. It is important to note that these definitions offer valuable perspectives on trust frameworks, focusing on agreements, standards, and regulatory conditions. However, no definition given is exhaustive. Five out of eight respondents did not give a concise definition of the term.

Components. All of the respondents named components from all four previously identified categories: legal, governance, operational and technical. The following new components were identified, and were found important by multiple interviewees:

- Future-proof technology (respondent 2 and 3)
- Financial liability (respondent 1, 2, 4, 5, 6, 7 and 8)
- Code of conduct (respondent 1, 2, 5, 7 and 8)
- Trust services (respondent 1, 3 and 7)

The following was mentioned by only one respondent:

- Data stewardship (respondent 7)

We add these components to the components found in the literature in Fig. 1 to create a comprehensive overview.

Another interesting finding from the interviews is that two respondents (2, 5) stressed that a trust framework should contain technical guidelines, but no implementations. This is significantly different from what the literature regards as trust frameworks, where the majority of the papers focused on the technical implementation as the core of the trust framework.

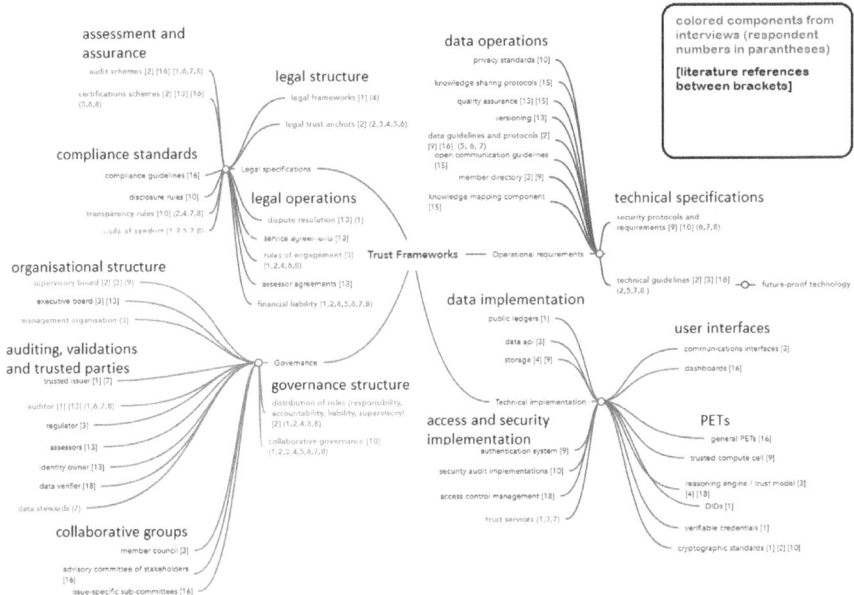

Fig. 1. Components of trust frameworks in the literature and from interviews

Taking these results into account: often the respondents do not necessarily agree with the literature, and some components of trust frameworks are only mentioned in one paper or by one respondent. We can therefore consider the components described in Fig. 1 as exhaustive but not exclusive to the components of trust frameworks. Table 4 describes the subcategories that these components are divided in.

Goals. The goal of trust frameworks came forward more strongly in the interviews, and can be summarized into six main categories: efficiency, interoperability, certainty, security, economic viability, societal impact.

Efficiency

All respondents mentioned efficiency as a goal of trust frameworks in one way or another. Respondent 3 mentioned that the standards created by trust frameworks are necessary for large-scale innovation:

> "Trust frameworks are essential, and generally digitalization consists of trust frameworks. (...) If we do not make agreements on how video works, then my camera would not work with both Zoom and WebEx. So if you want to work on innovation and create large-scale innovation in a chain, forms of standards will emerge".

Efficiency also extends to regulatory compliance, where trust frameworks could streamline this process and reduce complexities of compliance. By adhering to standardized protocols, organizations can navigate regulatory requirements more effectively, saving time and resources while ensuring adherence to legal obligations. Respondent 5 mentioned that trust frameworks generally require fewer resources to create a solution for everyone, after which it is widely implementable. This type of widely implementable standard will most likely emerge from Europe or the US and Big Tech according to one respondent, but it is not useful to wait until this happens:

> "The Netherlands does not decide the world standard, not even Europe. We could wait for an American or Big Tech standard, but there is no use to that. [So right now] we are looking at how we can make it better in the Netherlands".

Four respondents (1, 4, 5, 8) mentioned that there currently are enough technical solutions to make this large-scale innovation work, but they are not used on a large scale as of now. Trust frameworks could use these techniques for increased quality and efficiency of information chains.

Interoperability

Interoperability, or more specifically syntactic interoperability, is the application-level interoperability that allows multiple software components to cooperate even though their implementation languages, interfaces, and execution platforms are different [17]. Respondent 8 exemplifies how oftentimes when we use terms to describe data we are not exactly talking about the same concept:

"[It is important] to use the same language. For example, if you would like to know someone's income, what do you mean? Is that the fiscal income? There are a lot of definitions for the term. That could lead to difficult situations, because something might have been meant differently then it is interpreted, and this results in poor information exchange".

This interoperability could be achieved through the adaptive standardisation that a trust framework provides. Two respondents (6,7) also mentioned that this interoperability could reduce vendor lock-in within domains. Trust frameworks describe the requirement for services and the communication between them. Several parties can join and more suppliers could emerge around the same service. The trust framework would assure that these communicate in the same manner, making the barrier for switching between vendors minimal.

Certainty
Certainty of data and documents, as well as legal certainty have operational and societal implications. All respondents mentioned that the quality of data, the certainty of data transfers or the legal certainty of documents would improve when standards are in place to make set conditions for this data. Respondent 6 exemplifies improved legal certainty:

"It is mostly about the certainty that [for example] an accountant would have: 'This [document] was actually sent by the municipality or the tax authorisation, so I can use this in the process with certainty'"

Moreover, respondent 7 mentions that in an increasingly digitalized world where face-to-face interactions are limited, trust frameworks can play a role in enabling certainty and trust without physical contact, mitigating risks associated with digitalized or remote engagements.

Security
All respondents mentioned security as a goal of trust frameworks in one way or another. Online security could increase for businesses as well as individuals. Respondent 6 mentioned that trust frameworks could help create an official digital address making sensitive interaction far more safe:

"Currently there is no [all-purpose] digital address for individuals. Email addresses are not official digital addresses. (...) If I would like to work digitally, I need a digital address where the tax authorisation, government, private organisations, etc can reach me in a trusted and secure manner".

The increase of digital business will also increase fraud opportunities in online networks. When state-of-the-art technical standards are widely used, and implemented in a standardised way, the risk for fraud could decrease. Personal data will in turn also be shared and kept more securely. Fraud in digital environments happens in numerous ways, respondent 2 illustrates this with how we currently sign digitally:

"You give someone something very personal, namely your signature. This might not be perceived as very personal, but it is. In that organisation [which you send your autograph to], you have no idea what people could do with it. Someone could save it, everyone can see it"

Respondent 7 speaks of creating a comfort zone where parties can trust each other through security measures and agreements, and data can be shared safely:

"What might be the most important part of trust frameworks (...) is the creation of a certain comfort zone where people can share data. (...) It is often about data that can be very sensitive, or where decisions are being made with a societal impact. The comfort zone is the security: will we not lose the data? What is the source? Is there a risk for a man-in-the-middle attack? Using security, you earn trust".

Overall trust frameworks would model trust and security more similarly to how it is modelled in the physical world, where trust is created through agreements and laws. As illustrated by respondent 8:

"When you get into a public bus, you probably do not think 'I should check the brakes of this bus.'. Sometimes it is logical that we do not think about this anymore. In our digital world [this is different]: when you install an app or use a service, there is often be an entire book's worth of policy on how your data will be processed, and you have to find out legally what is written and what is actually meant. [You could compare this] with finding out by yourself whether the brakes work before using the app or service. Many people will therefore click 'accept' because few people will have the time, willingness or knowledge to [micromanage their trust]".

Economic Viability
Six respondents (1, 2, 3, 5, 6, 7) mention some form of economic gain in the use of trust frameworks. If many parties use one system, the costs will go down. Furthermore, the increased digitalization of data could reduce administrative costs, according to respondent 1:

"A lot of documents [in healthcare] are being retyped. The standards and the way of work in TIP could be a blessing for hospitals to reduce their costs and solve the problem [of re-entering information]".

Respondent 7 names the reductions of administrative costs that results from the decrease of paperwork. Moreover, respondent 2 names that the increased ease of doing online business could also give a boost to the internet and online business.

Three respondents (1, 2, 7) mention that mutual relationships and the creation of new business relations is another goal for trust frameworks. The enhanced trust fosters stronger partnerships and collaborations, leading to more efficient and productive business interactions. The frameworks also lower the

Understanding Trust Frameworks 235

Table 4. Components of trust frameworks

category name	description	example
Legal specifications		
assessment and assurance	Processes and schemes focused on ensuring compliance and standards through audits and certifications	Audit schemes, certification schemes
compliance standards	Guidelines, rules, and codes established to ensure legal compliance within the organizational framework	Compliance guidelines, transparency rules
legal structure	Frameworks and anchors that define the legal basis and requirements for operations	Legal frameworks, legal trust anchors
legal operations	Operational aspects related to legal matters	Dispute resolution, service agreements
Governance		
organisational structure	The structure and roles within the organization, including various boards and management entities	Supervisory board, executive board
auditing, validations and trusted parties	Responsibilities and roles of entities involved in auditing, validation, and secure issuing	Trusted issuer, auditor, regulator
collaborative groups	Groups established for collaborative purposes within governance	Member council, advisory committee
governance structure	Structure for governance, focusing on the distribution of roles	Distribution of roles, collaborative governance
Operational requirements		
data operations	Operations related to managing data	Privacy standards, knowledge sharing protocols
technical specifications	Specifications detailing the technical requirements, protocols, and guidelines for systems	Security protocols, technical guidelines
Technical implementations		
data implementation	Implementation aspects concerning technical data management	Public ledgers, data APIs
access and security implementation	Implementation aspects related to access control, security measures, and trust services	Authentication systems, security audit implementations
user interfaces	Interfaces designed for user interaction with systems	Communications interfaces, dashboards
PETs	Technologies aimed at enhancing privacy, including models, standards, and trusted environments	General PETs, trusted compute cell

barriers of entry for businesses seeking to enter new markets or engage with new partners. Through interoperable standards and protocols, businesses can connect with a broader network of suppliers, customers, and stakeholders. These relationships and the network creates opportunities for knowledge sharing. By promoting open collaboration and information exchange, trust frameworks facilitate the sharing of insights, expertise, and resources across domains. Through collaborative platforms, businesses can learn from each other's experiences, leverage emerging technologies, and drive continuous improvement and innovation.

Societal Impact
All respondents agreed that societal causes could benefit from the implementation of trust frameworks. The following examples were named that could be partially alleviated by trust frameworks:

- irrefutability of documents and identities is not sufficiently guaranteed (2, 6, 7)
- leaking or losing of important information with societal impact (7)
- unfair distribution of costs and benefits in business (4, 7)
- lack of certainty for citizens due to processing times of public organisations (7)
- closed business format and lack of inclusivity (7, 8)
- high administrative load on healthcare workers (1)
- lack of open, honest and transparent processes (2, 4, 7, 8)
- citizens being excluded to means e.g. people who do not have access to financial means (2)

However, respondent 5 argues that societal causes will never be the main goal for trust frameworks, as the incentive will always come through some operational or financial goal.

5 Conclusions and Future Research

This explorative paper provides a comprehensive conceptualisation of trust frameworks, including components and goals. Drawing on literature and the case study, we can answer the three research questions as follows.

Considering the first research question (what are the goals actors strive for with trust frameworks?), we found that the literature broadly outlines goals within six categories: security and privacy, certainty and compliance, societal impact, interoperability, efficiency, and trust. The goals emerging from the case study can be categorized into six categories: efficiency, interoperability, certainty, economic viability and societal impact. Out of these goals, all respondents agreed on four of them: interoperability, certainty, efficiency and security. All four of these goals also emerge from the literature. While the other goals were not agreed upon to be a universal goal of trust frameworks.

Considering the second research question and (which components are developed for achieving these goals?), we found four categories of components: technical specifications, governance, operational requirements, and legal requirements.

However, even though technical specifications are widely used as a component in the literature, not all respondents agree that this should be a part of trust frameworks. Similarly, even though the categories of components can be mostly agreed upon, a fair amount of the components that we found are only mentioned in one article or by one respondent, indicating even more that there is no consensus on the essential components of trust frameworks. Therefore, no minimal set of components for trust frameworks has been found.

The research, while explorative, lays the foundation for future research into trust frameworks. Trust frameworks represent a major change in traditional approaches to cross-sector data exchange and could use further research and development. By investigating trust frameworks across diverse use cases, exploring their transformative potential, and drawing upon insights from analogous concepts, we can progress the academic literature and provide real-world solutions for complex problems. Future research should expand to multiple use cases, to explore trust frameworks in various sectors and contexts for a more complete view. Additionally, drawing parallels with analogous concepts from related fields may offer novel perspectives and methodologies that can help in the development of trust frameworks.

References

1. Abramson, W., Hall, A.J., Papadopoulos, P., Pitropakis, N., Buchanan, W.J.: A distributed trust framework for privacy-preserving machine learning. In: Gritzalis, S., Weippl, E.R., Kotsis, G., Tjoa, A.M., Khalil, I. (eds.) TrustBus 2020. LNCS, vol. 12395, pp. 205–220. Springer, Cham (2020). https://doi.org/10.1007/978-3-030-58986-8_14
2. Bharosa, N.: The rise of GovTech: trojan horse or blessing in disguise? a research agenda. Government Inf. Q. **39** (2022). https://doi.org/10.1016/j.giq.2022.101692
3. Brewer, S., et al.: A trust framework for digital food systems. Nat. Food **2**, 543–545 (2021). https://doi.org/10.1038/s43016-021-00346-1
4. Cardoso, R.C., Gomes, A.: Towards a trust framework for multi-user virtual environments. In: Murgante, B., et al. (eds.) ICCSA 2014. LNCS, vol. 8579, pp. 754–768. Springer, Cham (2014). https://doi.org/10.1007/978-3-319-09144-0_52
5. Cho, J.H., Chan, K., Adali, S.: A survey on trust modeling. ACM Comput. Surv. (CSUR) **48**(2), 1–40 (2015)
6. Das, A.: Developing dynamic digital capabilities in micro-multinationals through platform ecosystems: assessing the role of trust in algorithmic smart contracts. J. Int. Entrep. **21**, 157–179 (2023). https://doi.org/10.1007/s10843-023-00332-7
7. Dijkhuis, S., Van Wijk, R., Dorhout, H., Bharosa, N.: When willeke can get rid of paperwork: a lean infrastructure for qualified information exchange based on trusted identities. In: Proceedings of the 19th Annual International Conference on Digital Government Research: Governance in the Data Age, pp. 1–10 (2018)
8. Gerybaite, A.: Digital governance: the case of proofs of vaccination, pp. 450–454 (2021). https://doi.org/10.1145/3494193.3494254
9. Getha-Taylor, H.: Cross-sector understanding and trust. Public Perform. Manage. Rev. **36**(2), 216–229 (2012)

10. Hardjono, T., Deegan, P., Clippinger, J.: Social use cases for the ID3 open mustard seed platform. IEEE Technol. Soc. Mag. **33**, 48–54 (2014). https://doi.org/10.1109/MTS.2014.2345197
11. Iroju, O., Soriyan, A., Gambo, I., Olaleke, J., et al.: Interoperability in healthcare: benefits, challenges and resolutions. Int. J. Innov. Appl. Stud. **3**(1), 262–270 (2013)
12. Kharvi, P.: Security risks, user privacy risks, and a trust framework for the metaverse space, pp. 119–123 (2023). https://doi.org/10.1109/MetaCom57706.2023.00033
13. Kusiak, L.: Baas over eigen zorgdata. Zorgvisie ICT **19**(4), 12–14 (2018)
14. Lu, B., Zhang, T., Wang, L., Keller, L.: Trust antecedents, trust and online microsourcing adoption: an empirical study from the resource perspective. Decis. Support Syst. **85**, 104–114 (2016). https://doi.org/10.1016/j.dss.2016.03.004
15. More, S.: Trust scheme interoperability: connecting heterogeneous trust schemes. In: Proceedings of the 18th International Conference on Availability, Reliability and Security, pp. 1–9 (2023)
16. Mpofu, N., Staden, W.V.V.: A trust framework model for identity-management-as-a-service (idmaas), pp. 455–462 (2015)
17. Ram, S., Park, J., Lee, D.: Digital libraries for the next millennium: challenges and research directions. Inf. Syst. Front. **1**, 75–94 (1999)
18. Ramsheva, Y., Prosman, E., Wæhrens, B.: Dare to make investments in industrial symbiosis? A conceptual framework and research agenda for developing trust. J. Clean. Prod. **223**, 989–997 (2019). https://doi.org/10.1016/j.jclepro.2019.03.180
19. Raturi, A., et al.: Cultivating trust in technology-mediated sustainable agricultural research. Agron. J. **114**, 2669–2680 (2022). https://doi.org/10.1002/agj2.20974
20. Release, E.P.: New interoperable Europe act to deliver more efficient public services through improved cooperation between national administrations on data exchanges and it solutions (2022). https://ec.europa.eu/commission/presscorner/detail/en/ip_22_6907. Accessed 6 Mar 2024
21. Resnick, P., Sami, R.: Sybilproof transitive trust protocols. In: Proceedings of the 10th ACM Conference on Electronic Commerce, pp. 345–354 (2009)
22. Rouhani, S., Deters, R.: Data trust framework using blockchain technology and adaptive transaction validation. IEEE Access **9**, 90379–90391 (2021). https://doi.org/10.1109/ACCESS.2021.3091327
23. Statista: Volume of data/information created, captured, copied, and consumed worldwide from 2010 to 2020, with forecasts from 2021 to 2025 (2021). https://www-statista-com.tudelft.idm.oclc.org/statistics/871513/worldwide-data-created/. Accessed 15 Mar 2024
24. Temoshok, D., Temoshok, D., Abruzzi, C.: Developing trust frameworks to support identity federations. US Department of Commerce, National Institute of Standards and Technology (2018)
25. Trusted Information Partners: Eenvoudig en betrouwbaar online zakendoen (2024). https://www.trustedinformationpartners.nl/. Accessed 19 Mar 2024
26. European Union: Regulation (EU) no 910/2014 of the European parliament and of the council of 23 July 2014 on electronic identification and trust services for electronic transactions in the internal market and repealing directive 1999/93/EC (2014). https://eur-lex.europa.eu/eli/reg/2014/910/oj. Accessed 12 Mar 2024
27. Yin, R.K.: Case Study Research: Design and Methods, vol. 5. Sage (2009)

Author Index

A
Alaerts, Luc 163
Ansah, Dwayne 179

B
Becker, Jörg 47
Berntzen, Lasse 116
Bharosa, Nitesh 223
Brützke, Paul 47

C
Christis, Maarten 163
Crompvoets, Joep 64

D
de Andrade, Luiz Henrique Alonso 31
Di Maria, Marco 47
Dijkhuis, Sander 223
Distel, Bettina 47

E
Edelmann, Noella 116
Effing, Robin 1

F
Francey, Alizée 147, 209

H
Hayes, Thomas 99
Heggertveit, Ida 16, 31
Hinz, Michael 1
Hofmann, Sara 31

J
Janssen, Marijn 223
Johannessen, Marius Rohde 116

K
Karlsson, Martin 99
Koddebusch, Michael 47

M
Mettler, Tobias 147, 209
Mori, Leonardo 147, 209

N
Nguyen, Binh An Patrick 194

P
Pauwels, Michiel 163

R
Reich, René 163
Roehl, Ulrik B. U. 64
Rydén, Hanne Höglund 31

S
Sattlegger, Antonia 131
Scholta, Hendrik 194
Spante, Maria 83
Stepanovic, Stefan 209
Susha, Iryna 179

V
van Acker, Karel 163
van der Peet, Louise 223
Vercalsteren, An 163

W
Wihlborg, Elin 83

SPRINGER NATURE

GPSR Compliance

The European Union's (EU) General Product Safety Regulation (GPSR) is a set of rules that requires consumer products to be safe and our obligations to ensure this.

If you have any concerns about our products, you can contact us on ProductSafety@springernature.com

In case Publisher is established outside the EU, the EU authorized representative is:

Springer Nature Customer Service Center GmbH
Europaplatz 3
69115 Heidelberg, Germany

The manufacturer's authorised representative in the EU is Springer Nature Customer Service Centre GmbH, Europaplatz 3, 69115 Heidelberg, Germany. If you have any concerns regarding our products, please contact ProductSafety@springernature.com

Printed and bound by CPI Group (UK) Ltd, Croydon, CR0 4YY

26/03/2026

02078962-0006